# Signal Processing
# for Industrial Diagnostics

# WILEY SERIES IN MEASUREMENT SCIENCE AND TECHNOLOGY

## Chief Editor

**Peter H. Sydenham**
*Australian Centre for Test & Evaluation*
*University of South Australia*

---

# Signal Processing
# for Industrial Diagnostics

**T. M. Romberg**
*IET Consulting Pty Ltd*
*Australia*

**J. L. Black** (Deceased)
**T. J. Ledwidge**
*University of Southern Queensland*
*Australia*

JOHN WILEY & SONS
Chichester • New York • Brisbane • Toronto • Singapore

*Other Wiley Editorial Offices*

John Wiley & Sons, Inc., 605 Third Avenue,
New York, NY 10158-0012, USA

Jacaranda Wiley Ltd, 33 Park Road, Milton,
Queensland 4064, Australia

John Wiley & Sons (Canada) Ltd, 22 Worcester Road,
Rexdale, Ontario M9W 1L1, Canada

John Wiley & Sons (Asia) Pte Ltd, 2 Clementi Loop #02-01,
Jin Xing Distripark, Singapore 0512

*Library Cataloging in Publication Data*
Romberg, T. M.
    Signal processing for industrial diagnostics / T.M.Romberg, J. L.
Black, T.J.Ledwidge
    p.  cm. – (Wiley series in measurement science and
technology
    Includes bibliographical references and index.
    ISBN 0 471 96166 3
    1. Detectors – Industrial applications.   2. Signal processing –
-Industrial applications.   I. Black, J. L., Ph.D.   II. Ledwidge, T.
J.   III. Title.   IV. Series.
TA165.R66 1996
670.42 – dc20.                                    96–5957
                                                        CIP

*British Library Cataloguing in Publication Data*

A catalogue record for this book is available from the British Library

ISBN 0 471 96166 3

Produced from camera ready copy supplied by the authors using Word 6.0
Printed and bound in Great Britain by Bookcraft (Bath) Ltd.
This book is printed on acid-free paper responsibly manufactured from sustainable
forestation, for which at least two trees are planted for each one used for paper production.

# DEDICATION

*This book is dedicated in memory of our co-author, Associate Professor Jim Black, who was tragically killed in an accident in the latter stages of drafting the manuscript for this book. His memory will live on with his friends, colleagues and, in particular, his students, with whom he was very popular. His co-authors mourn the loss of his lively and stimulating mind, and hope that his contribution to industrial signal processing, seen in part through this manuscript, will be recognised as a fitting gift from him to the wider community.*

# CONTENTS

## PART ONE: OVERVIEW AND THEORETICAL CONCEPTS

## PART TWO: INDUSTRIAL CASE STUDIES

# Series Editor's Preface

The series provides authoritative books, written by internationally acclaimed experts, on the topics that constitute measurement science (today also known as sensing) and its engineering.

Measurement data from system sensors needs complex processing to yield sound decisions for system management and control.

This title, the combined work of three skilled and experienced authors, is distinctly different from other signal processing works in that the target is industrial plant applications where diagnostics are the key aspects of interest. The many examples and case studies given here also set it aside as a fine teaching text and as a reference work for those applying these techniques to systems found in practice.

**Peter Sydenham**
Editor in Chief

# ACKNOWLEDGEMENTS

The emphasis of this book is on the practical applications of signal processing for industrial diagnostics. The authors have found that, when teaching courses on signal processing, the theory was made relevant by reference to our experience in applying the principles in real life situations.

Many of our students, in choosing topics in signal processing for their final year degree project, later completed doctoral studies in this area. One example is Simon Rofe who, as a co-author of Industrial Case Studies A and B, worked on the International Atomic Energy Agency research project from which the two case studies are drawn.

Other case studies cited in the book are also drawn from research projects completed with colleagues and numerous collaborators in industry. The contributions of all concerned are gratefully acknowledged. Our colleague, Dr Bob Harris, deserves a special mention in this regard. His influence on numerous collaborative projects over many years is evidenced in some of the case studies.

Associate Professor Nigel Hancock provided constructive criticism of some of the Chapters, and suggested some structural changes which improved the presentation. The Series Editor, Professor Peter Sydenham, also provided valuable encouragement and helpful comments on the first draft manuscript.

The Editorial and Production staff at John Wiley also deserve special mention for their helpful advice, editorial comment and their prompt response to numerous faxes.

Finally, but by no means least, we wish to thank our respective wives: Diane, who supplied numerous cups of coffee during the long hours of hard labour, and Jean, who read many of the Chapters and suggested changes which made the text clearer.

**T. M. Romberg & T. J. Ledwidge**
*February 1996*

# THE AUTHORS

### Dr T. M. Romberg
*BSc(Tech), BE, MEngSc, PhD, MIEAust, CPEng, MACS*

**Dr Romberg** is Managing Director of IET Consulting Pty Ltd, which specialises in strategic technology management and consulting. He holds tertiary qualifications in mechanical, nuclear and electrical engineering, and, prior to his present appointment, was Professor and Dean of the former School of Information Technology at the University of Southern Queensland and Project Manager and Principal Research Scientist in the CSIRO. He has extensive industrial and research experience in Australia, the U.K. and continental Europe in the steel and power industries, including the nuclear power industry in the U.K. and the Australian Atomic Energy Commission. Dr Romberg's research interests are in the fields of digital signal processing, mathematical modelling and advanced process control, and he is the inventor of the COALTROL™ advanced control software for the coal industry. He is the author of numerous scientific publications and has made invited visits to China and the U.S.A. as a keynote speaker in these fields of research.

### Associate Professor J. L. Black
*BSc(Hons), MSc, PhD, MIEAust*

**Dr Black** commenced his professional career in the mining industry in the U.K., and worked with the South African Atomic Energy Authority prior to taking up an academic appointment as a lecturer at the Capricornia College of Advanced Education in Rockhampton, Queensland in the early 1970s. He returned to the U.K. for a further period in the mining industry, and then emigrated to Australia in 1975 to take up an appointment as a lecturer at the Darling Downs Institute of Advanced Education. He was promoted to senior lecturer then Associate Professor for his contribution to teaching and research. He developed courses in control engineering, computer systems and signal processing. His many research interests included computer controlled systems as well as the IAEA funded LMFBR project with Professor Ledwidge. He was the Chief Scientific Investigator for this project when he was tragically killed in an accident.

### Emeritus Professor T. J. Ledwidge
*Dip Elec Eng (Dist), BSc (Hons), PhD, FIEAust, CPEng*

**Professor Ledwidge** had wide experience in electrical manufacturing, the coal industry, the nuclear industry and academia. His nuclear engineering experience included a period as Principal Scientific Officer at the world's first liquid metal cooled fast breeder reactor (LMFBR) at Dounreay in Scotland. He emigrated to Australia in 1970 to take up an appointment as Head of Engineering Physics and Principal Research Scientist with the Australian Atomic Energy Commission. In 1974, he was appointed Foundation Dean of Engineering, then Pro-Vice-Chancellor (Academic) and finally, by invitation in 1990, to the position of Vice-Chancellor of the University College of Southern Queensland, which he led through its crucial transitional stages to full University status as the University of Southern Queensland in 1992. Although retired, Professor Ledwidge has continued to teach signal processing to engineering and science undergraduates and to supervise PhD students assisting with his continuing research in nuclear engineering. This research on the detection of incipient malfunction of LMFBRs has been funded annually since 1985 by the International Atomic Energy Agency (IAEA) in Vienna.

# PART ONE

# OVERVIEW
# AND
# THEORETICAL
# CONCEPTS

# 1

## OVERVIEW AND PRINCIPLES

### 1.1    RATIONALE

From the dawn of time, the human race has been measuring the world and the universe in which we live. Whether it was the weather, the seasons of the year, the phases of the moon or the time of day, there has always been an insatiable desire to measure and analyse 'what is there'. The range of instruments and devices for measuring our world and the mega-universe beyond has grown enormously during the twentieth century. These enable us to investigate everything from the smallest particles and microscopic creatures (electron microscope) to vast constellations in space (Hubble telescope). This phenomenal growth in instrumentation technology will continue to accelerate as we move to discover new frontiers of science in the twenty-first century.

Recent advances in measurement systems stem largely from the significant improvements in instrumentation technology and the corresponding rapid developments in computer and digital signal processing technology, together with the availability of sophisticated software analysis tools. Practicing engineers and scientists are confronted with a greater need for comprehensive information to monitor the integrity of their complex processes, and are often confused by the ever increasing range of 'user friendly' software packages which promise ever more elegant analysis and visual presentation of their process data. However, a real danger with all this progress and change is the indiscriminate application of sophisticated analysis tools without the requisite depth of knowledge needed to interpret the results obtained.

The *principle aim* of this book is to provide the practitioner in the field of industrial diagnostics with a comprehensive understanding of the fundamental signal processing techniques that underpin modern digital diagnostic systems, and to demonstrate how they are applied through practical examples and industrial case studies. In teaching these techniques to students over many years, the authors have concluded that the rapidly growing field of *information science*, and its subset *signal processing* in particular, needs to be taught to graduates and undergraduates as a coherent discipline from a *process applications* viewpoint. This approach is in stark contrast to most electrical engineering and mathematical texts on this subject, which tend to target the burgeoning communications field.

*Signal processing* as a discipline involves the measurement and analysis of fluctuations about the mean values of selected process variables, and may be performed with one or more of the following objectives in mind:

- the detection of an incipient malfunction (*process diagnostics*);
- the adjustment of selected variables for controlling and/or optimising operation (*process control*); or
- the estimation of physical quantities or phenomena such as flow velocity, mixing length or natural frequency of vibration (*process identification*).

These objectives usually necessitate continuous monitoring and analysis of signals from a process within well defined, often very short, time scales in order to provide useful information. The *science* and *technology* of signal processing are discussed in the following sections.

### 1.1.1    The Science of Signal Processing

Signal processing is regarded here as a subset of information science in recognition of the fact that the end point is to extract useful information about an industrial process. 'Information' is used here in the same way as it is used by Shannon [1948, 1949] and by Wiener [1948] as 'that which removes uncertainty'. The notion of uncertainty relies on the basic ideas of probability analysis developed axiomatically, for example, by Papoulis [1991] and is now defined in terms of probabilistic descriptions by the International Organisation for Standardisation [1993].

Shannon's pioneering work was essentially concerned with the process of communication in the presence of noise in order to establish limits on the rate of transmission and the consideration of the generation of possible errors in the transmission process. Modern communication engineering continues this theme which is made more complex by the demands of contemporary society. Considerable attention is currently focused on the application of signal processing strategies within this communication engineering context.

The emphasis here is on the application of appropriate signal processing strategies for the solution of industrial problems. Therefore, it parallels much of the work in communication engineering, financial modelling, medical diagnostics and other processes where the measured parameters are essentially unpredictable. It shares the same mathematical framework and uses the same functions to describe stochastic processes. The mathematical treatment in the book is intended to highlight the application in industrial diagnostics rather than develop an axiomatic approach more suitable to a course in mathematical statistics. It is hoped that part two of the book, dealing with practical case studies, will assist in the desired objective.

### 1.1.2    The Technology of Signal Processing

When the authors first worked in industry using some of the signal processing techniques discussed in this book, the technology available was very primitive by today's standards. At that time, over one-third of a century ago, signals were multiplied together using Hall effect multipliers while a delay between two signals was implemented using recordings on a magnetic tape which was then cut and spliced into a continuous loop to facilitate analysis over a range of narrow frequency bands in a sequential manner. To generate an estimate of even one simple frequency spectrum would take almost one whole day! This is in stark contrast with the technology available today where most standard spreadsheets have built-in analysis packs to calculate spectral and correlation estimates in a matter of a few minutes.

In those early years the authors used one inch magnetic tape on aluminium drums to store data for later analysis. In our current research we receive experimental data on standard size CDs each holding about 600 Mb of data. A modestly priced desktop computer is all that is usually required to perform reasonably comprehensive analysis of data contained on the CDs.

While the burden of analysing data has almost disappeared, the onus of making a correct diagnosis still remains. Because analysis can now be accomplished so easily the temptation to produce elegant graphs and pictures, rather than concentrate on diagnostic techniques, is great. As mentioned earlier there is a real danger in this approach and the authors hope that the readers will be assisted in recognising when one of the techniques outlined in this book is relevant to their problem (and also, of course, when it is not).

### 1.2    DEFINITIONS

### 1.2.1    General Metrological Terms

The general metrological terms used in this book are those given in the international vocabulary of basic and general terms in metrology, published by the International Organisation for Standardisation (ISO).

Following the convention in the ISO guide, in this book the same symbol is used for the physical quantity (the measurand) and for the random variable that represents the possible outcome of an observation of that quantity. This results in considerable economy of notation when the context of the discussion makes the usage clear.

### 1.2.2    Deterministic, Random and Stochastic Processes

Processes[1], whether natural or artificial, and the measurements or signals which are used to characterise their transient or dynamic behaviour, can be divided into one of two classes: deterministic processes or random processes.

*Deterministic processes* (and signals) are those which can be described precisely by a mathematical expression for example, a straight line or a trigonometric function.

*Random processes* (and signals, also termed random variables in many texts) are those which cannot be described precisely by mathematical expressions, and can only be characterised by statistical measures in the amplitude (or probability), time and frequency domains. The spatial coordinates of a leaf fluttering in the wind is an example of a random process.

*Stochastic processes* (and signals) are random processes and signals which satisfy convergence and limit theorems associated with the probabilistic and asymptotic properties of random sequences. The word stochastic is derived from the Greek and implies 'by trial and error' or 'hit and miss'.

Although random and systematic errors are now defined by the ISO in precise statistical terms, it is convenient in the present context to classify these errors together with other unknown flaws inherent in the mathematical modelling process as an extraneous or residual noise component superimposed on the measurable inputs and outputs of the process. Stochastic and random processes are analysed using the same techniques; therefore, we will often use these terms interchangeably in the text. The reader should note, however, that the official terms are not synonymous.

### 1.2.3    Time Variant, Invariant and Stationary Processes

Consider the signal measured on a random process as shown in Figure 1.1. The signal $x(t)$ is varying randomly as a function of time $t$, and is said to be time variant. Its mean value (or time average) $\overline{X} = \lim_{T\to\infty} \int_0^T x(t)dt$, however, is independent of time as shown by the dashed line, and is said to be time invariant. If a series of sample records $x_j(t)$ ($j = 1$ to $N$) of the same signal are recorded at subsequent times and the mean values $\overline{X}_j$ of each sample record are equal (i.e. $\overline{X}_1 = \overline{X}_2 = .... = \overline{X}_N$), the process is said to be *stationary*.

### 1.2.4    Linear and Nonlinear Processes

A process is *linear* if it obeys the principle of superposition. For example, a

---

1.  The term *processes* (*i.e.* operations or changes) rather than the term *systems* (*i.e.* sets of interconnected parts) will be used throughout this book.

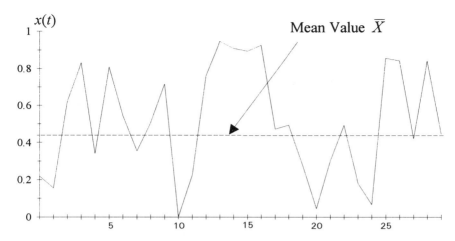

**Figure 1.1** Typical random (or stochastic) signal

process with output $y_1(t)$ is functionally related to its input $x_1(t)$ by the equation, $y_1(t) = f(x_1,t)$, and it is superimposed on to the second process $y_2(t) = f(x_2,t)$ such that the resulting process $y(t)$, Figure 1.2, is given by

$$y(t) = y_1(t) + y_2(t) = f(x_1,t) + f(x_2,t) = f(x_1 + x_2,t)$$

thus both process 1 and process 2 and the resulting process are *linear* by definition. (We will make extensive use of this property in the following chapters.) However, if the above superposition relationship does not hold, that is, if

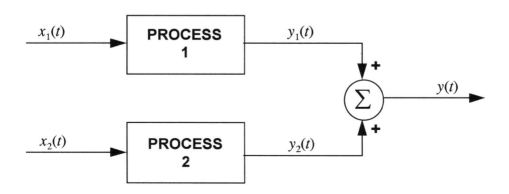

**Figure 1.2** Typical process with two inputs and one output

$$y(t) \neq y_1(t) + y_2(t)$$

then process 1 and/or process 2 *and* the resulting process are *nonlinear*.

### 1.2.5    Univariate, Bivariate and Multivariate Processes

A *univariate process* is one which is characterised by a *single* measurement (signal). In the example of Figure 1.2 the output $y(t)$ of the composite process may be the only measurement which can be made in practice, and certain assumptions need to be made about the number and the properties of the non-measurable input(s) in order to deduce its impulse or frequency response characteristics. Univariate processes are discussed in Chapter 2.

A *bivariate process* is one which is characterised by two measurements (signals). In the example, process 1 would be characterised by its input and output measurements, $x_1(t)$ and $y_1(t)$ respectively, and process 2 by its respective input and output, $x_2(t)$ and $y_2(t)$. However, it is more common in practice to measure only one input, (say) $x_1(t)$, and the output $y(t)$, and to ignore the influence of the second (and any other processes) that may be present; that is, those variables not measured are relegated to a non-measurable, extraneous noise source superimposed on the output. Bivariate processes are discussed in some detail in Chapter 4, and form the basic building block for higher order multivariate processes.

A *multivariate process* is one which is characterised by three or more measurements (signals). In the above example, Figure 1.2, the multivariate process would be characterised by measurements of the inputs, $x_1(t)$ and $x_2(t)$, and the output $y(t)$. In general, however, a multivariate process may consist of an arbitrary number of input and output measurements, and the matrix of possible interactions between measurands characterising the multivariate process are discussed in Chapter 7.

Most currently available signal processing software packages can only perform *bivariate analysis* on the signals monitored on a process. Since industrial processes are predominantly multivariate in their dynamic behaviour, a major focus of this book is on the multivariate analysis of process signals.

## 1.3    PRINCIPLES OF INDUSTRIAL DIAGNOSTICS

### 1.3.1    A Definition of Industrial Diagnostics

The practical data analysis exercises discussed in some detail in chapter eight, and all of the case studies reported in chapter nine, have a common goal. This goal is to make definitive statements about the process (or state of the plant).

These statements exhibit knowledge about the process that is often only available by use of the signal processing strategies used in these applications.

Thus our definition of industrial diagnostics is *the interpretation of information gleaned from sensors in order to assess the condition of a process or to provide an estimate of a process variable*. For example, in the data analysis exercises we demonstrate how cross-correlation is used to measure the average velocity of a stream of hot gas and how the mixing length may be estimated, and in the industrial case studies we show how spectral analysis is used to detect incipient boiling in a liquid metal cooled fast reactor.

## 1.3.2    Stages of Industrial Diagnostics

There are four steps in the implementation of any industrial diagnostic strategy and these are

- selection of appropriate sensors[2].
- extraction of features from the sensors.
- comparison of features with defined standards.
- determination of a reliable decision process.

Each of these is discussed in turn in the following sections, but it should be born in mind that in any particular application the user may not have complete control over all four steps. Often the decision to implement a diagnostic scheme is taken after the plant is operational and, in this case, compromises are often necessary. The most common problem encountered by practitioners is that the sensors, which are already installed for operational purposes, are often unsuitable for diagnostic studies or they are not installed in the best location. In these circumstances the remaining three steps are critical and should be approached with caution.

### 1.3.2.1    The Selection of Appropriate Sensors

A major consideration in selecting the type of sensor to be used in any particular application is an insight into the underlying science of the process. This insight is exploited to generate signals from the process with as high a signal as possible and with minimum disturbance from other sources. It should be recognised at the outset that sensors designed to respond to one particular physical variable are often influenced to some extent by other variables. For example a simple strain gauge used to detect changes in stress applied to the wire will also respond to changes in temperature. The same gauge will also

---

2.    The terms transducer, pick-off and detector are often used interchangeably with sensor. The term sensor is used throughout this book.

produce a random noise signal due to the thermal agitation of its constituent molecules. These effects cannot be eliminated completely, but may be brought within tolerable bounds by careful choice of sensor and consideration of the communication channel through which the information is transmitted.

Taking, as an example, detection of localised boiling in a liquid-metal-cooled fast reactor, the use of thermocouples that are responsive to changes in temperature in the liquid metal seems to be an obvious first choice. However, a limitation of this choice is the fact that fluctuations in temperature, which accompany incipient boiling, tend to be localised and convected along with the flowing liquid metal. This requires that temperature sensors would have to be installed at the outlet of each fuel sub-assembly. Although this approach is acceptable the utilisation of phenomena with global characteristics removes the physical constraint associated with the need to place sensors in specific geometrical positions.

Acoustic noise generated by the boiling process and perturbations in nuclear reactor flux caused by voids in the liquid metal are both phenomena which may be detected at a distance from the localised boiling site.

Acoustic noise is generated by the rapid growth and collapse of vapour bubbles when the liquid metal boils and is propagated with very little attenuation throughout the whole of the reactor structure. In this case simple accelerometers attached to the outside of the reactor vessel result in a situation which is easy to maintain, producing signals with characteristics that may be defined following suitable calibration tests.

Neutrons in a reactor are often treated as though they form a continuous fluid, instead of the discrete particles they actually are. The discrete nature of the neutrons and the statistical nature of the fission process that is a pre-cursor to their existence generate random fluctuations in the neutron population which is often called reactor noise. This noise is modified by absorption and moderation of other components of the reactor system as the neutrons spread out from their origin. In a liquid metal cooled fast reactor, voids in the sodium caused by localised boiling will influence the reactor noise and, in principle will be detectable remote from the site of localised boiling by the use of the current from an ionisation chamber.

The transmission of the energy from the site of the localised boiling is beyond the control of the user but needs to be considered in any diagnostic strategy. For example, acoustic signals will excite some parts of the mechanical structure into resonance and the reactor flux will be changed by passage through areas of varying cross-section. Lack of understanding of these effects may lead to incorrect diagnosis of incipient fault conditions or perhaps to designing the feature extraction scheme on false premises.

It is vitally important to pay appropriate attention to the communication channel through which the signals from the sensors are processed. Although the extraction of specific features from the signals is of paramount importance it is usual to use as much *a-priori* knowledge as is available in order to enhance the signal to noise ratio as much as possible. These insights, gained from experience, are generally more valuable than those relying solely on the unthinking application of advanced signal processing strategies. Combining this insight with comprehensive signal processing strategies can often turn the detection of an 'anomalous event' into meaningful information about the process under examination.

Having decided on the types of sensors suitable for the detection of relevant phenomena, it is important to recognise that the sensors, and parts of the communication channel, are not always entirely reliable. This aspect of accommodating faults in the detection system is traditionally achieved by distributing sensors and associated hardware components around the plant in order to insure against damage in any one localised channel. Such arrangements are typically configured in a triplex or quadruplex arrangement with the output from each channel compared for logical consistency, with small variations between them being ignored. Such an approach is often called hardware redundancy although the concept is often extended to include analytical processes, in which case parallel redundancy is a more appropriate term.

When building hardware redundancy into a system it is vital to remember that identical sensors tend to have similar life expectancies, and hence a malfunction in one sensor is likely to be closely followed by faults in others. This limitation is overcome by using dissimilar sensors responsive to different process variables. Returning once again to the example of detecting incipient boiling in a liquid metal fast reactor, we conclude that a combination of sensors responding to acoustic noise, variations in nuclear flux and fluctuations in temperature would achieve optimum parallel redundancy.

### 1.3.2.2   Extraction of Features from the Sensors

An important application of industrial diagnostics is the early detection of the onset of an anomaly. We take this application to introduce the concept of extraction of features. In simple terms the problem may be represented as

$$z(t) = x(t) + \delta * y(t)$$

where $z(t)$ is a typical signal available from the sensor, and $x(t)$ and $y(t)$ are the signals associated with the background noise in the plant and the signal arising from the onset of an anomaly respectively. The operator $\delta$ has a value of zero when there is no anomaly present and unity otherwise. The task in detecting the occurrence of an anomaly is simply to estimate when the operator is sufficiently removed from zero and close enough to unity to constitute the indication of the onset of an anomaly. We shall see later that in practice this

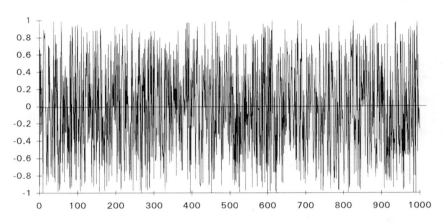

**Figure 1.3** A typical sample of background noise in an industrial plant

involves a decision making process, as discussed in sections 1.3.2.3 and 1.3.2.4.

At this stage it is necessary to use whatever *a priori* knowledge or creative insights that may be available to help in establishing a suitable feature extraction strategy. In the present exercise the knowledge that the test signal consists of a sinusoidal burst is used by exploiting the orthogonal properties of sinusoids by multiplying the signal by a sinusoid of the same frequency and averaging over a number of samples.

For purposes of illustration, one feature extraction strategy is demonstrated by using a simulation model. In this model actual background noise from an industrial plant is mixed with a test signal which is switched on for a defined time. The test signal has a mean square level considerably below that of the background noise but is known to be a sinusoid of a given frequency. The object of the simulation is to detect the times when the test signal is switched on and off.

The normal background noise in the industrial plant is shown in Figure 1.3 over 1000 data samples. The vertical axis is arbitrary units but is to the same scale as that used for the test signal which is shown in Figure 1.4. Note that the maximum value of the test signal is one-quarter that of the maximum value of the background noise.

The composite function is now formed by the addition of the background noise with the test signal to produce the result shown in Figure 1.5. It should be noted that even in this case where the test signal is very well defined, an examination of the composite signal reveals neither the onset nor the duration of the defined anomalous occurrence. This indicates that we need to extract some feature from the composite signal to assist in the decision making process. The result of this is shown in Figure 1.6.

The decision boundary is set sufficiently high above the background noise

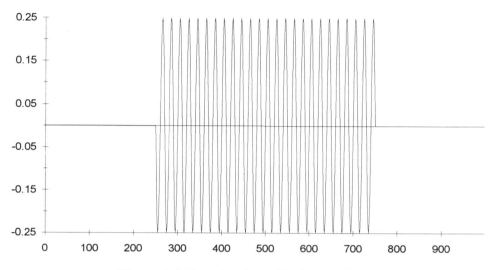

**Figure 1.4** Test signal used in the simulation

level to make the occurrence of spurious indications within acceptable limits, but not so high that legitimate anomalies remain undetected. These aspects are examined further in section 1.3.2.4.

The above example is a simple application of feature extraction when some key aspect of the anomaly is known. Other more sophisticated techniques are required when precise knowledge of the anomalous process is unknown. In the example described in section 1.3.2.2, it was reasonably deduced that the test anomaly occurred as indicated in Figure 1.6 by comparing the output of the

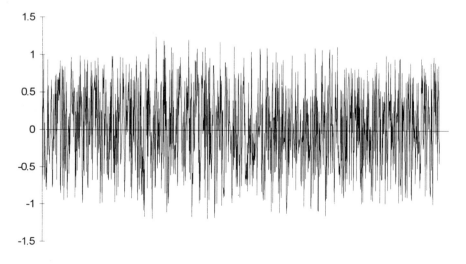

**Figure 1.5** The addition of background noise and test 'anomaly'

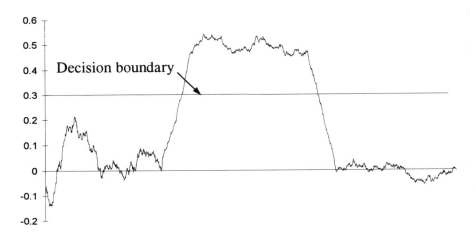

**Figure 1.6** Feature extracted by spectral analysis averaged over 100 data samples

spectral filter over sample periods when the industrial plant was operating in 'normal' conditions. It is usual to collect sample records over significantly long periods of operation in order to establish a base line defined as normal operation. The standards derived in this way are often called historical standards.

Historical standards are widely used in industrial applications because they are relatively easy to establish although, as we shall see later in section 1.3.2.4, uncertainties always arise in the decision making process. A major reservation with the use of these standards is the lack of precise data on the unknown. Some of these feature extraction strategies are explored in Chapters 9 and 10.

### 1.3.2.3   Comparison of Features with Defined Standards

In the example described in section 1.3.2 it was reasonably deduced that the test anomaly occurred as indicated in Figure 1.6 by comparing the output of the spectral filter over sample periods when the industrial plant was operating in 'normal' conditions. It is usual to collect sample records over significantly long periods of operation in order to establish a base line performance feature extraction strategies during an actual departure from normality. Although 'normal' operation is the goal of operators of industrial plant, confidence in the fault detection scheme would no doubt be enhanced by additional perspective on the long term statistical nature of base line data.

One method that is used in some situations is to test prototypes of key elements of the plant for the whole of their life expectancy in order to generate signals arising during failure of the element and establish base line data over very long periods. Standards established by prototype testing generally have wide acceptance amongst operators of industrial plant.

One of the practical data analysis exercises included in chapter nine uses information generated in the simulation of fault conditions in a heat exchanger. This simulation is as real as possible in order to provide a calibration standard for use in practice. Calibration by direct methods is usually difficult to perform, but when available it generates considerable confidence in the fault detection system.

In some simple situations it is possible to establish a comprehensive theoretical standard by rigorous mathematical analysis. Although examples of this approach are hard to find in practice one such application is included in chapter eight to illustrate the value of theoretical standards when they are available.

### 1.3.2.4    Determination of a Reliable Decision Process

The ultimate aim in any fault detection system is to provide an unambiguous indication of the onset of fault conditions as soon as possible and do so with utmost reliability. The criteria for assessing the quality of the decision include:

- the time $T$ taken before an indication of fault conditions is reached;
- the probability $P_s$ of generating a spurious indication of a fault;
- the probability $P_m$ of missing the onset of fault conditions.

These three criteria are all interrelated, which necessitates a consideration of the trade-offs involved in designing any particular application.

Suppose that the extraction of a feature from a plant operating under 'normal' conditions results in a Gaussian probability density function (pdf)[3] with zero mean value. The onset of a fault is assumed to shift the mean value of the feature to a new non-zero value. These probability distribution functions are shown in Figure 1.7 which also shows a decision boundary, arbitrarily set at the intercept of the two probability distribution functions.

It is to be expected that as the averaging time $T$, over which the features are estimated, is increased the overlap between the two pdfs decreases. $T$ of course determines the speed of response of the diagnostic system to a fault condition. The choice of $T$ is thus determined by the maximum time allowable in any practical situation. In plant with a high risk of severe damage as a result of the escalation of an undetected fault such as a nuclear reactor, $T$ would be defined to be around one-half of one second.

---

3.    Probability density functions are discussed in detail in Chapter 2.

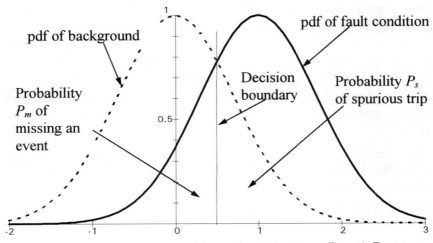

**Figure 1.7** Illustration of the trade-off between $P_s$ and $P_m$

The automatic detection of a fault requires the setting of a decision boundary which when crossed in feature space initiates the indication of the fault condition. When only one feature is used, this corresponds to a single threshold, as shown in Figure 1.6. An examination of this Figure reveals that the choice of the decision boundary is also a compromise, as either $P_s$ or $P_m$ can be made as small as we please, but only at the expense of the other. The actual position of the decision boundary is usually set with reference to defined operational criteria. For example, some power generation authorities will tolerate the spurious indication of fault conditions no more than once per year, but set a considerably longer time associated with the probability of missing the onset of a fault. The implementation of these constraints requires knowledge of the probability density functions of signal features related to normal and fault conditions respectively.

The reliability of the decision making process can be improved by using more than one feature extracted from the signal, providing of course that the features are independent. A simple example of multiple features would be the simultaneous use of an amplitude probability function with the power within a defined frequency band. We could then set optimal thresholds for each feature and specify decision rules ranging from the exceeding of both thresholds simultaneously to either threshold being exceeded, the former having the same philosophical base as parallel redundancy discussed earlier in section 1.3.2.1. In practice the setting of a decision surface is a complex task when several signal features are used. The gains, however, can be substantial and well worth the effort and added complexity of the computation required.

## 1.4 STRUCTURE OF THE BOOK

The presentation structure aims to provide a balance between theory and practice so that the book will serve as a reference text for both students and practicing professionals. For this purpose, the book is divided into two parts. Part 1 contains the theoretical concepts necessary for a complete understanding of the practical topics discussed in Part 2, which contains the data analysis exercises and industrial case studies. Each chapter in Part 1 starts with a set of learning objectives listing the topics to be discussed. Readers are advised to study the objectives and attempt the problems at the end of each chapter. Professionals familiar with the topics may elect to review the chapter or move on to a chapter more relevant to their immediate needs.

**Chapter 1** gives an overview and introduces the terminology used in addition to setting out the rationale of the book. The principles of industrial diagnostics are established for use later in the book.

**Chapter 2** sets out the important theoretical relationships which underpin the analysis techniques discussed in later chapters. The signals monitored in industrial processes are predominantly random in nature, and their statistical (probability domain) properties are discussed in conjunction with the concept of ergodicity to facilitate transformation from the probability to the time and frequency domains. The time domain correlation functions and frequency domain spectral functions form Fourier transform pairs known as the Wiener-Khinchin equations.

**Chapter 3** presents the more important concepts of sampled data processes and the signal sampling and pre-processing operations that need to be adopted in practice in order to obtain reliable estimates of the time and frequency domain functions. The topics include: sampling theorem, Z-transforms, discrete processes, the Fast Fourier Transform (FFT), signal conditioning and the estimation of functions from finite length records.

**Chapter 4** analyses single-input-single-output (SISO) bivariate processes, and presents a proof for calculating reliable process transfer function estimates in the minimum least squares sense. It discusses the effects of uncorrelated noise, data filtering, data smoothing using lag or spectral windows, and process time delays. The book includes MS Windows-compatible computer programs for SISO simulations to demonstrate the underlying principles. Bivariate processes are pivotal to the analysis of higher order multivariate processes discussed in the following chapter.

**Chapter 5** looks at the case where the measurements on a process are made inside a closed loop, that is, one in which the input measurements are influenced by return feeds or controller action from the process output(s). It presents a set of criteria for identifying closed loop processes from input-output measurements, and discusses the use of dither signals.

**Chapter 6** reviews more recent parametric spectral analysis methods which are being widely used in many ares of science and engineering, particularly the geophysical sciences. The parametric methods attempt to maximise the 'entropy' of the signal in order to 'optimise' the order of an autoregressive prediction error filter, and equivalence relations with the previous non-parametric methods are derived.

**Chapter 7** discusses both multiple-input-multiple-output (MIMO) and multiple-input-single-output (MISO) multivariate processes, and presents a proof for reducing MIMO processes to a set of independent MISO sub-processes, the MISO partial spectral density, transfer function and coherence estimates of which are computed from the augmented matrix of ordinary spectral density estimates using a recursive algorithm developed by one of the authors. The book also includes MS Windows-compatible computer programs for MISO simulations to demonstrate the underlying process analysis and computational principles.

**Practical Data Analysis Exercises** are provided on computer discs accompanying this volume, and entail laboratory simulations of practical problems previously solved by the authors in industrial situations. The authors firmly believe that the most effective way to demystify the inherent mathematical concepts is to use an applications approach wherever possible. This observation has been confirmed many times through teaching engineering and science students at both undergraduate and graduate levels. The use of data taken from real life gives the subject a relevance that stimulates students and has been selected to cover a range of processes, including:

- the estimation of the time constant of a thermocouple
- the estimation of the natural frequency and damping coefficient of a vibrating cantilever excited by a random input
- the measurement of the average velocity of a column of hot gas and estimation of the turbulent mixing length
- detection of the onset of a sodium to water reaction in a heat exchanger of a nuclear liquid metal cooled fast breeder reactor
- the location of a noise source in a dispersive medium.

Students are encouraged to attempt these exercises in order to understand the analysis principles involved in each case study. The exercise on a vibrating cantilever, in particular, has proved to be a useful vehicle for helping students master some aspects of advanced mathematics through experiencing the practical benefits of analysing the random signals involved in the process. In our experience, students are amazed to observe how useful information about a randomly vibrating cantilever can be extracted with elegance from seemingly incoherent signals.

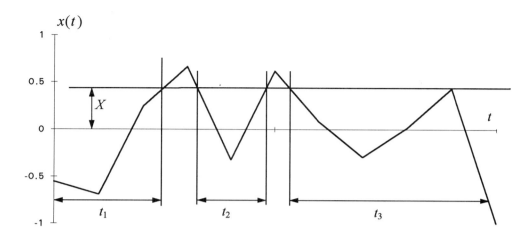

**Figure 2.1** Typical random signal illustrating the time spent below a defined amplitude level $X$

## 2.2.1 The Amplitude Probability Function

The amplitude probability function is a measure of the probability of a signal being below a given amplitude level at a given instant of time. For an *ergodic process* (defined later), it is the fraction of the total sample length that a signal spends below a given amplitude level. This is illustrated in Figure 2.1.

If $F(X)$ is the probability that the signal $x(t)\leq$ the amplitude $X$, and is defined as

$$F(X) = \lim_{T\to\infty} \frac{t_1 + t_2 + t_3 + \ldots + t_{n-1} + t_n}{T},\qquad \ldots (2.1)$$

then it follows that

$$F(\infty) = 1$$
$$F(-\infty) = 0$$

$F(0) = 0.5$ for a signal of zero mean value.

The general form of the amplitude probability function $F(X)$ is given in Figure 2.2. Note that the function is always asymptotic to unity as the value of $X$ increases towards infinity. If the signal has zero mean value, then the amplitude probability function passes through the point $(0, 0.5)$ signifying that the signal spends as much time above the mean value as it does below it. Note also that if the signal is continuous and differentiable, then the amplitude probability function will be smooth as shown in this example. Even if the signal

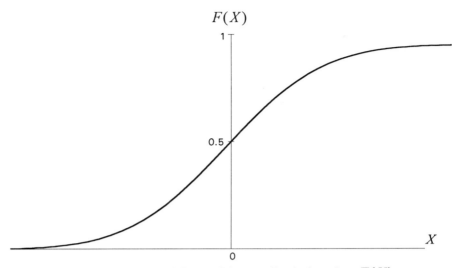

**Figure 2.2** General form of the amplitude function $F(X)$

is such that it leads to a probability function which is not smooth, the probability function is always asymptotic to unity and always passes through the point (0,0.5) for a signal with a zero mean value. The reader may care to compute and plot the amplitude probability function for a square wave to illustrate this point.

As a specific example, consider the calculation of the amplitude probability function $F(X)$ associated with a sine wave over one period $(-T \rightarrow +T = 2T)$ given as

$$x(t) = A\sin(\omega t).$$

In this case the time spent below an amplitude of $X$ is shown in Figure 2.3.

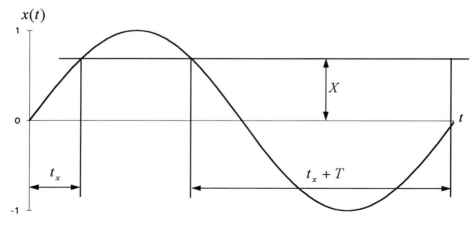

**Figure 2.3** Illustration of the time spent below an amplitude $X$ for a sine wave

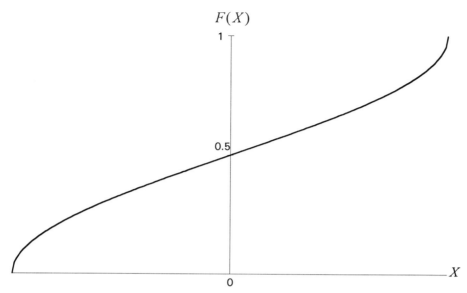

**Figure 2.4**  Amplitude probability function $F(X)$ for a sine wave

Using the definition of $F(X)$ and, with reference to Figure 2.3, this results in

$$F(X) = (2t_x + T)/2T,$$

where, with $\omega = \dfrac{\pi}{T}$, $t_x$ is given as $t_x = \dfrac{T}{\pi} \sin^{-1}\left(\dfrac{X}{A}\right)$.

Substituting this expression for $t_x$, we have the following result:

$$F(X) = \left\{ \frac{2T}{\pi} \sin^{-1}\left(\frac{X}{A}\right) + T \right\} / 2T = \frac{1}{\pi} \sin^{-1}\left(\frac{X}{A}\right) + \frac{1}{2}. \qquad \text{... (2.2)}$$

This function is shown in graphical form in Figure 2.4.

### 2.2.2    Probability Density Function

In the previous section it was noted that when the probability of a signal is equal to or less than a given value, $X$ is defined by the amplitude probability function $F(X)$. In this section the situation where the signal lies within a narrow window around the neighbourhood of $X$ is examined, noting that the probability that:

$$x(t) \leq (X + \delta X) \text{ is } F(X + \delta X)$$

and the probability that

$$x(t) \leq X \text{ is } F(X).$$

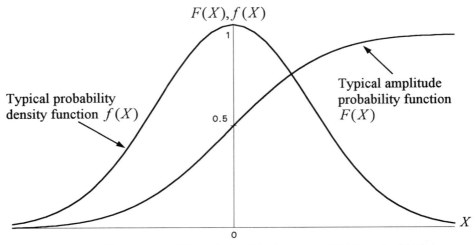

**Figure 2.5** Illustration of the relationship between $F(X)$ and $f(X)$

Hence the probability that $x(t)$ lies between $X$ and $(X + \delta X)$ is given as:

$$\text{Prob}\{X \leq x(t) \leq X + \delta X\} = F(X + \delta X) - F(X)$$

$$\text{Prob}\{X \leq x(t) \leq X + \delta X\} = \left\{\frac{F(X + \delta X) - F(X)}{\delta X}\right\}.\delta X$$

$$\text{Prob}\{X \leq x(t) \leq X + \delta X\} = \{F'(X)\}.\delta X .$$

The **probability density function** (pdf) is defined as the derivative of the amplitude probability function, as follows:

$$f(X) = \frac{dF(X)}{dX}, \qquad \text{... (2.3)}$$

where the relationship between the pdf $f(X)$ and the amplitude probability function $F(X)$ is shown in Figure 2.5.

With reference to Figures 2.4 and 2.5 we note that the probability of the signal $x(t)$ lying between two values of $X$ may be expressed in one or other of the following forms.

In terms of the probability density function we have:

$$\text{Prob}\{X_1 \leq x(t) \leq X_2\} = \int_{X_1}^{X_2} f(X)dX$$

or, alternatively, in terms of the amplitude probability function we have:

$$\text{Prob}\{X_1 \leq x(t) \leq X_2\} = F(X_2) - F(X_1)$$

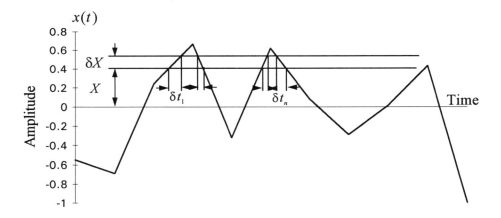

**Figure 2.6** Illustration of the meaning of the probability density function

Another way to represent the probability density function is to consider the time a signal spends within a narrow window as a fraction of the total time of the sample. This is illustrated in Figure 2.6 in which the width of the window is $\delta X$ and it is located at a height of $X$. The signal resides in the window for a series of short times $\delta t_1, \delta t_2, \delta t_3, ..., \delta t_n$ as shown in Figure 2.6. The total time spent within the window is simply the sum of these elemental times. The probability of finding the signal within the window is thus the fraction of the total sample length $T$. For an ergodic process this probability, obtained from an estimate of the fraction of time spent in the window, is equal to the product of the pdf with the width of the window as shown in equation (2.4).

$$\text{Prob}\{X \le x(t) \le (X + \delta X)\} = f(X).\delta X = \frac{\delta t_1 + \delta t_2 + ... \delta t_n}{T}$$

... (2.4)

$$f(X)\delta X = \sum_{i=1}^{n} \frac{\delta t_i}{T}$$

### 2.2.3 Expected Value Operator

Although a detailed description of a random variable in the amplitude domain is provided by its probability distribution, it is often convenient to use a concise description of this information. Such a description is provided by the *expected value*, which for a random variable that assumes a finite number of values is simply the average value of all the values weighted by their respective probabilities of occurrence. For example, if a random variable takes on values $X_1, X_2, ..., X_n$ with respective probabilities of $P_1, P_2, ..., P_n$ its expected or average value is given as

$$\text{Expected value} = X_1 P_1 + X_2 P_2 + \ldots X_n P_n .$$

In general, the following expression is used:

$$E(X) = \sum X_i P_i \qquad\qquad \ldots (2.5)$$

when the random variables are discrete and, when they are continuous, as

$$E(X) = \int_{-\infty}^{+\infty} xf(x)dx . \qquad\qquad \ldots (2.6)$$

The concept of expectation extends to functions of random variables. For example if a function of a random variable is expressed as $g(x)$ then the expectation of this function is

$$E[(g)X)] = \int_{-\infty}^{+\infty} g(x)f(x)dx . \qquad\qquad \ldots (2.7)$$

## 2.3      TIME DOMAIN ANALYSIS

A typical random signal obtained from one run of an experiment might look something like the record shown in Figure 2.7. This signal has a mean value, in this case around 0.4. It is usual to subtract the mean value from the signal to produce a signal of zero mean value for subsequent analysis.

It is usual to collect a number of sample records which is called an *ensemble*. Such an ensemble can be obtained by taking a very long recording of the signal under investigation and dividing it into an arbitrary number of records, or taking many individual records. A typical ensemble is shown in Figure 2.8.

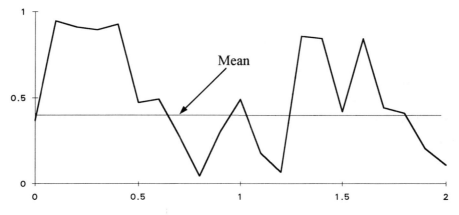

**Figure 2.7** A typical random signal with a mean value around 0.4

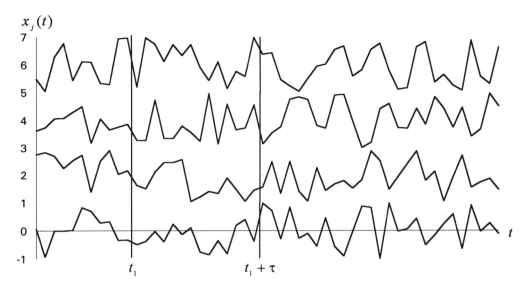

Figure 2.8   Illustration of an ensemble of records

## 2.3.1   Ensemble and Time Averages

From the collection of records there is the option of calculating the time average of each and every member of the ensemble or calculating the average across the ensemble at any specified time. Assuming the time at which an average is calculated is $t_1$, then the average across the ensemble (ensemble average) is given as equation (2.8)

$$\text{Ensemble average} = \lim_{N \to \infty} \frac{1}{N} \sum_{i=0}^{N} x_i(t_1) \qquad \dots (2.8)$$

If the ensemble average is independent of the choice of time origin the process (or signal) is called **stationary**; that is, a stationary process is one in which

$$\lim_{N \to \infty} \frac{1}{N} \sum_{i=0}^{N} x_i(t_1) = \lim_{N \to \infty} \frac{1}{N} \sum_{i=0}^{N} x_i(t_1 + \tau) \text{ for all values of } \tau. \quad \dots (2.9)$$

On the other hand, the average over time of the $i$th record is given as:

$$\text{Time average} = \lim_{T \to \infty} \frac{1}{T} \int_{0}^{T} x(t)dt . \qquad \dots (2.10)$$

If the ensemble average and the time average are equal to each other, then the process is termed **ergodic** which implies that the time average can be deduced from the ensemble average and vice-versa. This is an assumption commonly made in practical applications and allows the solution of problems that otherwise would be intractable. (See, for example, the problem at the end of this chapter dealing with the estimation of the mean square of a random triangular wave).

Stationarity and ergodicity both imply that the length of the record of the signal is infinitely long, thus requiring an approximation in practical applications.

### 2.3.2    Mean, Mean Square and Variance

The estimation of the mean is commonly calculated by using the following expression:

$$\text{Mean value } \overline{X} = \lim_{T\to\infty} \frac{1}{T} \int_0^T x(t)dt\,. \qquad \text{... (2.11)}$$

The mean square is calculated using the following relationship, again operating in the time domain.

$$\text{Mean square } \overline{X^2} = \lim_{T\to\infty} \frac{1}{T} \int_0^T x^2(t)dt\,. \qquad \text{... (2.12)}$$

An ergodic process may be approached from the time domain or the probability domain. In the latter case note that the fraction of time a signal spends within a window $\delta X$ wide is simply $f(X)\delta X$. The contribution to the mean value of the signal is thus given by $Xf(X)\delta X$, resulting in the mean value as the sum of all such contributions as:

$$\text{Mean value } \overline{X} = \int_{-\infty}^{+\infty} Xf(X)dX\,. \qquad \text{... (2.13)}$$

By the same reasoning we may calculate the mean square as:

$$\text{Mean square } \overline{X^2} = \int_{-\infty}^{+\infty} X^2 f(X)dX\,. \qquad \text{... (2.14)}$$

### 2.3.3    Correlation Functions

The mean-square value of an ensemble is the average of the square of the function at a particular time. This concept may be extended by taking the average of two values of the variable separated by a time interval after an arbitrary time epoch. In this case the average of the 'lagged product' is defined as the **autocorrelation function** as follows:

$$R_{xx}(t_1,\lambda) = \lim_{N \to \infty} \frac{1}{N} \sum_{i=1}^{N} x_i(t_1)x_i(t_1+\lambda). \qquad ...(2.15)$$

If the process is **stationary** then the autocorrelation function is independent of time, and thus

$$R_{xx}(\lambda) = \lim_{N \to \infty} \frac{1}{N} \sum_{i=1}^{N} x_i(t)x_i(t+\lambda). \qquad ...(2.16)$$

Further, if the process is **ergodic** then we may use any member of the ensemble as representative and replace the ensemble by a time average, thus:

$$R_{xx}(\lambda) = \lim_{T \to \infty} \frac{1}{2T} \int_{-T}^{T} x(t)x(t+\lambda)dt. \qquad ...(2.17)$$

For the special case when the time lag $\lambda$ is zero the autocorrelation function becomes:

$$R_{xx}(0) = \lim_{T \to \infty} \frac{1}{2T} \int_{-T}^{T} x^2(t)dt. \qquad ...(2.18)$$

This value of $R_{xx}(0)$ is the **mean square** of the signal and is used to normalise the autocorrelation function to give the normalised autocorrelation, $\Phi_{xx}(\lambda)$. This function has a value of unity when the lag term $\lambda$ is zero; that is

$$\Phi_{xx}(\lambda) = \frac{R_{xx}(\lambda)}{R_{xx}(0)}. \qquad ...(2.19)$$

*General form of* $\Phi_{xx}(\lambda)$

A shorthand notation may be used for the autocorrelation function as

$$R_{xx}(\lambda) = < x(t)x(t+\lambda) >,$$

which may be interpreted as the average of the lagged product. With this nomenclature consider the inequality

$$< \{x(t) - x(t+\lambda)\}^2 > \geq 0.$$

which, upon expansion, becomes:

$$< \{x^2(t) - 2x(t)x(t+\lambda) + x^2(t+\lambda)\} > \geq 0.$$

For a stationary process the first and the third terms in the above expression are both equal to the mean square, whereas the middle term is twice the autocorrelation function. This leads to the conclusion that

$$R_{xx}(0) \geq R_{xx}(\lambda). \qquad ...(2.20)$$

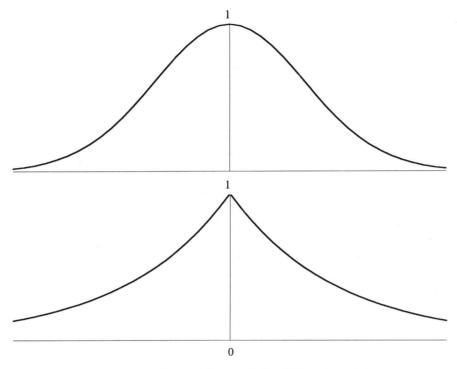

**Figure 2.9** General forms of $\Phi_{xx}(\lambda)$ at the origin

In other words, the autocorrelation function has its maximum value at the origin. Also note that for a stationary process the lag term may be taken as either positive or negative without affecting the value of the autocorrelation function which is thus symmetrical about the vertical axis and may be written:

$$\Phi_{xx}(\lambda) = \Phi_{xx}(-\lambda).\qquad\qquad\ldots(2.21)$$

This symmetry is confirmed by differentiating with respect to the lag variable $\lambda$ to give:

$$\Phi'_{xx}(\lambda) = -\Phi'_{xx}(-\lambda).\qquad\qquad\ldots(2.22)$$

Note in particular, that at the origin the above condition implies that the form of the autocorrelation function is either cusp-shaped or with a horizontal tangent as shown in Figure 2.9. If the mean value of the signal is removed prior to computing the autocorrelation function, the result is known as the **autocovariance function** and is defined as:

$$C_{xx}(\lambda) = E[\{x(t) - \bar{x}\}\{x(t+\lambda) - \bar{x}\}] = R_{xx}(\lambda) - \{\bar{x}\}^2.\qquad\ldots(2.23)$$

The concept of autocorrelation may be extended by considering the relationship between two variables If the average of the lagged product between a signal and the delayed version of another signal is calculated, the product is called the **cross-correlation function** and defined as

$$R_{xy}(\lambda) = \lim_{T \to \infty} \frac{1}{2T} \int_{-T}^{T} x(t)y(t+\lambda)dt \qquad \dots (2.24)$$

and

$$R_{yx}(\lambda) = \lim_{T \to \infty} \frac{1}{2T} \int_{-T}^{T} x(t+\lambda)y(t)dt . \qquad \dots (2.25)$$

Like the autocorrelation function, if $x(t)$ and $y(t)$ are stationary functions then

$$R_{xy}(\lambda) = R_{yx}(-\lambda). \qquad \dots (2.26)$$

however, unlike the autocorrelation function the cross-correlation function is not generally an even function, nor is it necessarily a maximum at the origin.

The degree of correlation between $x(t)$ and a delayed version of $y(t)$ is determined by a dimensionless correlation function defined as:

$$\Phi_{xy}(\lambda) = \frac{R_{xy}(\lambda)}{\sqrt{R_{xx}(0)R_{yy}(0)}}. \qquad \dots (2.27)$$

## 2.4    REVIEW OF CONTINUOUS FOURIER ANALYSIS

### 2.4.1    Fourier Series Representation of Continuous Signals

Any arbitrary function that satisfies the *Dirchlet* condition that $\int_{0}^{T} |x(t)|dt$ is finite may be expressed as a sum of cosinusoids and sinusoids, provided that the function is periodic the period being. If, for example, the function $x(t) = x(t+T)$ for all values of time t, then it may be written as follows:

$$x(t) = a_0 + a_1 \cos\omega_0 t + a_2 \cos2\omega_0 t + \dots + b_1 \sin\omega_0 t + b_2 \sin2\omega_0 t + \dots$$

or more succinctly as

$$x(t) = a_0 + \sum_{n=1}^{\infty} (a_n \cos(n\omega_0 t) + b_n \sin(n\omega_0 t)), \qquad \dots (2.28)$$

where the coefficients $a$ and $b$ are constants and

$$\omega_0 = \frac{2\pi}{T} \text{ radians} / s$$

$T =$ period in seconds

This representation of the periodic function $x(t)$ as a sum of sinusoids and cosinusoids is called a *Fourier series* and it has a *fundamental frequency* of $1/T$. Integral multiples of this fundamental frequency are called *harmonics*. To determine the coefficient $a_0$, integrate both sides of the above equation over one whole period, noting that the contribution from each sinusoid is zero, resulting in

$$a_0 = \frac{1}{T}\int_0^T x(t)dt .$$

To determine the general coefficient $a_n$, multiply each side of the above equation by $\cos n\omega_0 t$ and again integrate over one whole period, noting that the orthogonally property of sinusoids ensures that the only contribution is

$$\int \cos^2 n\omega_0 t dt = \frac{T}{2} .$$

Hence

$$a_n = \frac{2}{T}\int_0^T x(t)\cos n\omega_0 t dt \qquad\qquad \text{... (2.29)}$$

and, similarly,

$$b_n = \frac{2}{T}\int_0^T x(t)\sin n\omega_0 t dt . \qquad\qquad \text{... (2.30)}$$

### 2.4.2     Exponential Form of the Fourier Series

It is often convenient to use the exponential form of the Fourier series by substituting for the sinusoids and cosinusoids as follows:

$$\sin n\omega_0 t = \frac{e^{jn\omega_0 t} - e^{-jn\omega_0 t}}{2j}$$

$$\text{... (2.31)}$$

$$\cos n\omega_0 t = \frac{e^{jn\omega_0 t} + e^{-jn\omega_0 t}}{2}$$

Substituting these expressions into equation (2.27) (with some re-arrangement) gives:

$$x(t) = a_0 + \sum_{n=1}^{\infty}\frac{a_n - jb_n}{2}e^{jn\omega_0 t} + \frac{a_n + jb_n}{2}e^{-jn\omega_0 t} . \qquad \text{... (2.32)}$$

Now as all the cosine terms are *even* it follows that all the $a_n$ are *even* and hence $a_n = a_{-n}$. Similarly as all the sine terms are *odd* it follows that all the $b_n$

are *odd* and hence $b_n = -b_{-n}$. The above expression for $x(t)$ may then be written as

$$x(t) = a_0 + \sum_{n=1}^{\infty} \frac{a_n - jb_n}{2} e^{jn\omega_0 t} + \sum_{n=-1}^{-\infty} \frac{a_n - jb_n}{2} e^{jn\omega_0 t}. \qquad \text{... (2.33)}$$

Putting $C_n = \dfrac{a_n - jb_n}{2}$ and substituting for $a_n$ and $b_n$, the following results are obtained:

$$C_n = \frac{a_n - jb_n}{2} = \frac{1}{T} \int_0^T x(t)(\cos n\omega_0 t - j \sin n\omega_0 t)dt$$

$$C_n = \frac{1}{T} \int_0^T x(t) e^{-jn\omega_0 t} dt \qquad \text{... (2.34)}$$

from which we note by comparing the definition of $a_0$ that $C_0 = a_0$, resulting in the following compact expression for the periodic function $x(t)$:

$$x(t) = \sum_{n=-\infty}^{\infty} C_n e^{jn\omega_0 t} \qquad \text{... (2.35)}$$

In general, the coefficients $C_n$ are complex. The exponential form will be found to be convenient when developing the concept of power spectral density function in the following section.

## 2.5    FREQUENCY DOMAIN ANALYSIS

In section 2.3 it was shown that a random or stochastic signal may be conveniently summarised in the time domain by the use of the autocorrelation function and that two signals may be jointly analysed by the cross-correlation function. In many practical situations it is preferable to work in the frequency domain because often the visualisation of the process is more intuitively obvious. The functions used in the frequency domain are the **power spectral density** and the **cross-spectral density functions**, and are related to the time domain functions covered previously. These functions are explained in the following sections.

The technique of resolving a signal into its different frequency components is well known and widely used when the signals are deterministic, the general method of analysing these deterministic signals being the use of Fourier series. It is natural then to seek ways to extend the application of Fourier series to random or stochastic signals in order to describe these signals in the frequency domain.

The major topic examined in this section is the power spectral density function, which measures the distribution of the power of a signal in the frequency domain. It is a function that has considerable application in industrial diagnostics and is the subject of some of the case studies covered in part two of this book. The description of a signal in the frequency domain, rather than the time domain, does not increase the amount of information available and we should expect that these descriptions are related, as we shall see in section 2.6.

### 2.5.1    Power Density of a Pulse Train

By way of introduction consider the description of a deterministic signal in the frequency domain by using a Fourier series representation as introduced in section 2.4. The example chosen for this illustrative purpose is the signal from regular train of rectangular pulses, each pulse being of duration $\tau$ and spaced at regular intervals $T$ apart as shown in the Figure (2.10).

As the function is periodic, we may use the Fourier series representation defined by the following equations:

$$x(t) = \sum_{n=-\infty}^{\infty} C_n e^{jn\omega_0 t}$$

$$C_n = \frac{1}{T} \int_0^T x(t) e^{-jn\omega_0 t} \, dt \qquad \text{... (2.36)}$$

In this case the value of the function is $x(t) = A$ for $\dfrac{-\tau}{2} \le x(t) \le \dfrac{+\tau}{2}$, and hence

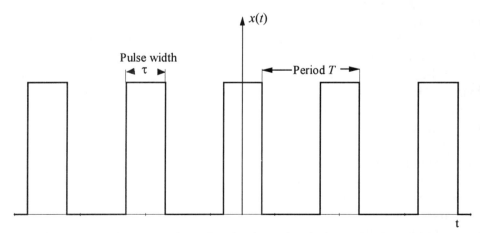

Figure 2.10  A rectangular train of pulses of period $T$ and pulse width $\tau$

$$C_n = \frac{1}{T} \int_{-\tau/2}^{+\tau/2} A e^{-j\omega_0 t} dt \qquad \ldots (2.37)$$

which, after integration and rearrangement becomes:

$$C_n = \frac{A\tau}{T} \text{sinc}\left(\frac{n\pi\tau}{T}\right). \qquad \ldots (2.38)$$

This 'line spectra' representation with the sinc function as its envelope is shown in Figure (2.11). The power (or mean square) in the $n$th component is simply

$$\frac{C_{-n}^2}{2} + \frac{C_n^2}{2} = C_n^2$$

by virtue of the symmetry of the line spectra, and may be imagined to be spread out over a frequency band $1/T$ wide. The power per unit bandwidth, or *power density*, is thus given as

$$\text{power density} = \frac{C_n^2}{\left(\frac{1}{T}\right)} \text{ with units of amplitude}^2 \text{ per Hz.}$$

Note that as the period tends to infinity (as would be the case for a random signal) the distance between the spectral lines tends to zero, thus forming a continuous spectral curve. This line of reasoning will be followed in developing the concept of power spectral density of a random signal in the next section.

### 2.5.2    Power Spectral Density of a Random Signal

In the previous section it was shown that Fourier series analysis may only be applied to signals that are periodic and thus are not applicable to random signals directly. This restriction is overcome by a strategy which generates an artificial periodic signal from a random signal by taking a section from the random signal and repeating that sample to generate a periodic signal. Fourier series analysis may then be applied to this artificial signal to generate the coefficients of the line spectra in the usual way. The strategy is then finalised by allowing the length of the sample taken from the random signal to increase without limit and examining the mathematical consequences.

If the original random signal is written as $x(t)$, then a sample of length $T$ from that signal may be denoted as $x_s(t)$ This sample signal is then repeated without limit to generate a periodic signal which may be called $x_p(t)$ and then Fourier series analysis may be used on this periodic signal as follows:

$$x_p(t) = \sum_{n=-\infty}^{\infty} C_n e^{j\omega_0 t}, \qquad \ldots (2.39)$$

where

$$\omega_0 = \frac{2\pi}{T}$$

and

$$C_n = \frac{1}{T} \int_{-T/2}^{T/2} x_p(t) e^{-j\omega_p t} dt .$$

The expressions may be made symmetrical by letting $X_n = C_n T$, in which case the following pair results:

$$x_p(t) = \sum_{n=-\infty}^{\infty} X_n e^{jn\omega_0 t} \frac{1}{T}$$

$$X_n = \int_{-T/2}^{+T/2} x_p(t) e^{-j\omega_0 t} dt$$

$$\qquad \text{... (2.40)}$$

The reasoning used in the previous section to determine the power spectral density of a rectangular train of pulses is followed by noting that the power in the $n$th component is $C_n^2$ which, by virtue of the substitution given above, is

$$C_n^2 = \left(\frac{X_n}{T}\right)^2 .$$

As before imagine this power spread out over a frequency band $1/T$ wide resulting in the power density being given as:

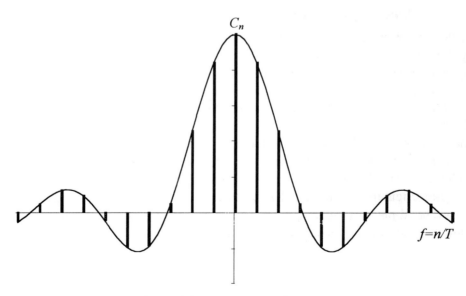

**Figure 2.11** Envelope of the line spectra of a rectangular pulse train

$$\text{power density} = \left(\frac{X_n}{T}\right)^2 / \frac{1}{T} = \frac{1}{T} X_n^2.$$

To recover the original signal let the length of the sample $T$ increase without limit and note that in so doing the following convergences occur

$$T \to \infty$$

$$\frac{1}{T} \to df$$

$$x_p(t) \to x(t)$$

$$n\omega_0 \to \omega$$

$$X_n \to X(\omega)$$

Substitution of the above into the equations of the Fourier series yields the **Fourier transform pair** as follows:

$$X(\omega) = \int_{-\infty}^{+\infty} x(t) e^{j\omega t} dt$$

$$x(t) = \int_{-\infty}^{+\infty} X(\omega) e^{-j\omega t} df$$

... (2.41)

The power density is given as

$$\frac{X_n^2}{T} \to \frac{1}{T} X(\omega) X^*(\omega),$$

where the superscript * denotes the complex conjugate. This leads to the *definition* of the **power spectral density** being

$$S_{XX}(\omega) = \lim_{T \to \infty} \frac{1}{T} \{ X(\omega) X^*(\omega) \}$$

... (2.42)

The power spectral density may be regarded as a representation of the way the power in a signal is spread out in the frequency domain from minus infinity $(-\infty)$ to plus infinity $(+\infty)$. The use of the concept of negative frequencies is simply to take advantage of the symmetrical nature of the double-sided Fourier transform. The real measurable power spectral density $G_{XX}(\omega)$ is related to the theoretical power spectral density as shown in Figure 2.12.

The total power (mean square) in the signal is the area under the curve and may be written in terms of the powers spectral density as

$$\text{mean square} = \frac{1}{2\pi} \int_{-\infty}^{+\infty} S_{XX}(\omega) d\omega,$$

... (2.43a)

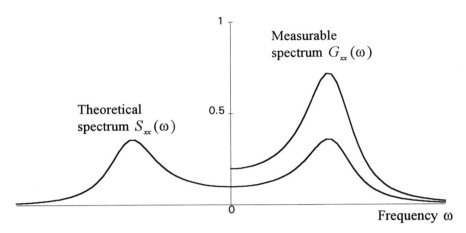

**Figure 2.12** Illustration of the relationship between the measurable and theoretical spectra

or in terms of the measurable spectrum as

$$\text{mean square} = \frac{1}{2\pi} \int_{-0}^{+\infty} G_{XX}(\omega)\,d\omega .$$   ... (2.43b)

If the power spectral density is constant over a very wide frequency range the spectrum is termed 'white', using the analogy of white light, although the latter has constant power per unit wavelength whereas white noise has constant power per unit frequency.

### 2.5.3    The Cross-Spectral Density Between Two Signals

Recall that in the previous section is was stated that the power spectral density is a representation of power in a signal summarised in the frequency domain. The *time domain* relationship between two signals has been discussed previously in this chapter by use of the *cross-correlation function*. The corresponding *frequency domain* relationship between two signals is given as the **cross-spectral density function**, and is defined in a manner similar to that of the power spectral density as

$$S_{XY}(\omega) = \lim_{T \to \infty} \frac{1}{T} \left\{ X^*(\omega) Y(\omega) \right\}$$   ... (2.44)

and

$$S_{YX}(\omega) = \lim_{T \to \infty} \frac{1}{T} \left\{ Y^*(\omega) X(\omega) \right\}.$$   ... (2.45)

In general the cross-spectral density function is a complex function with real and imaginary components, and thus may be written as

$$S_{XY}(\omega) = C_{XY}(\omega) - jQ_{XY}(\omega), \qquad \qquad \text{... (2.46)}$$

where the functions on the right hand side are the *co-spectral density* and the *quad-spectral density* functions respectively, and are defined alternatively as:

$$\begin{aligned} C_{XY}(\omega) &= \Re\{S_{XY}(\omega)\} \\ Q_{XY}(\omega) &= \Im\{S_{XY}(\omega)\} \end{aligned} \qquad \qquad \text{... (2.47)}$$

The next section shows that the spectral density functions and the correlation functions are, in fact, Fourier transform pairs.

## 2.6     WIENER-KHINCHIN EQUATIONS

In the case of periodic deterministic signals, the signals may be described either in terms of periodicity (in the time domain) or in terms of their line spectra (in the frequency domain). No new information is gleaned by moving from one domain to the other. Practical considerations usually determine which representation is the most useful in specific applications.

In dealing with stochastic signals it was seen that the appropriate measures by which they may be summarised are the correlation and spectral density functions. Wiener and Khinchin first drew attention to the fact that correlation and spectral density functions form Fourier transform pairs. This observation is illustrated in the case of the autocorrelation function and the power spectral density function as shown below:

The definitions of the autocorrelation function and the power spectral density function over samples $2T$ long give

$$R_{xx}(\lambda) = \lim_{T \to \infty} \frac{1}{2T} \int_{-T}^{+T} x(t)x(t+\lambda)dt$$

$$\qquad \qquad \text{... (2.48)}$$

$$S_{XX}(\omega) = \lim_{T \to \infty} \frac{1}{2T}\{X(\omega)X^*(\omega)\}$$

and taking the Fourier transform (FT) of the autocorrelation function gives

$$FT\{R_{xx}(\lambda)\} = \int_{-\infty}^{+\infty}\left\{\lim_{T \to \infty} \frac{1}{2T} \int_{-T}^{+T} x(t)x(t+\lambda)dt\right\}e^{-j\omega\lambda}d\lambda. \qquad \text{... (2.49)}$$

Substituting $\mu$ for $(t+\lambda)$ and noting that $d\lambda$ becomes $d\mu$, then

$$FT\{R_{xx}(\lambda)\} = \int_{-\infty}^{+\infty}\left\{\lim_{T \to \infty} \frac{1}{2T} \int_{-T}^{+T} x(t)x(\mu)dt\right\}e^{-j\omega(\mu-t)}d\mu \qquad \text{... (2.50)}$$

reults which, after expanding the exponential term and separating the two integrals, becomes

$$FT\{R_{xx}(\lambda)\} = \lim_{T\to\infty}\frac{1}{2T}\left\{\int_{-T}^{+T}x(t)e^{j\omega t}dt\int_{-\infty}^{+\infty}x(\mu)e^{-j\omega\mu}d\mu\right\}. \quad ... (2.51)$$

It is recognised that the two integrals represent the Fourier transform of x(t), i.e. $X(\omega)$ and its conjugate complex $X^*(\omega)$, and thus:

$$FT\{R_{xx}(\lambda)\} = \lim_{T\to\infty}\frac{1}{2T}X(\omega)X^*(\omega) = S_{XX}(\omega). \quad ... (2.52)$$

By identical reasoning it can be shown that the cross-correlation function and the cross-spectral density function also form Fourier transform pairs. The Wiener-Khinchin equations are usually written formally as

$$S_{XX}(\omega) = \int_{-\infty}^{+\infty}R_{xx}(\lambda)e^{-j\omega\lambda}d\lambda$$

$$\quad ... (2.53)$$

$$R_{xx}(\lambda) = \frac{1}{2\pi}\int_{-\infty}^{+\infty}S_{XX}(\omega)e^{j\omega\lambda}d\omega$$

These two equations exhibit the fact that there is no new information generated by moving from the time domain to the frequency domain or vice-versa. However, in some applications of industrial diagnostics it is easier to interpret the underlying phenomena in one domain or the other. The reader should examine the case studies in part two of this book to see some typical examples of this aspect.

One use of the first of the two equations given in equation (2.53) is to use it to calculate the power spectral density from a known estimate of the autocorrelation function. It is now more common to use the Fast Fourier Transform (FFT) to calculate the power spectrum directly from the raw time series associated with the signal under investigation. Note that the power spectral density functions detailed in the case studies were computed using one of the popular FFT algorithms available.

## TUTORIAL PROBLEMS

### Problems on Probability Density Functions

1) A random variable has the following probability density function (pdf) $f(X) = 1$ for $0 \le X \le 1$ and zero elsewhere. Calculate its mean and mean square.

2) Show that the pdf of $x(t) = A\sin(\omega t)$ is

$$f(X) = \frac{1}{\pi\sqrt{A^2 - X^2}}.$$

and, using this pdf, calculate the mean and mean square of the signal.

3) Show that the pdf of a square wave of amplitude $\pm A$ is

$$f(X) = \frac{1}{2}\delta(X - A) + \frac{1}{2}\delta(X + A).$$

Using this pdf calculate the mean and mean square.

4) A random binary signal occupies two states, $\alpha$ and $\beta$, spending twice as long at the first state as at the second one. Calculate the variance of the signal.

5) If the pdf of a signal is given as

$$f(X) = ae^{-2b|X|}$$

what is the relationship between $a$ and $b$?

6) The pdf of a Gaussian signal is given as:

$$f(X) = Ae^{-\left(\frac{X^2}{2\sigma^2}\right)}.$$

What is the value of $A$?

7). A random triangular wave of amplitude $\pm A$ is passed through a limiter which clips the signal at $\pm B$, where $B$ is less than $A$. Calculate the variance of the signal.

**Problems on Correlation Functions**

8) A signal is represented by the following expression:

$$x(t) = A \sin(\omega t + 9)$$

Derive an expression for the autocorrelation function.

9) Two stationary random signals, $x(t)$ and $y(t)$ are added to produce another random signal $z(t)$. Determine an expression for the autocorrelation function of $z(t)$:

a) when $x(t)$ and $y(t)$ are correlated, and

b) when $x(t)$ and $y(t)$ are uncorrelated.

10) A telegraph signal switches between $+A$ and $-A$ under the governance of a Poisson distribution designated as

$$P(k) = \frac{[\mu\lambda]^k}{k!}e^{-\mu\lambda}$$

where $P(k)$ is the probability of there being $k$ changes of sign in time $\lambda$ and $\mu$ is the average number of zero crossings per unit time. Calculate $R_{xx}(\lambda)$.

11) A random telegraph signal is added to a sine wave with an amplitude one-tenth of that of the telegraph signal. Derive an expression for the

autocorrelation function of the composite signal and plot this function to illustrate how periodic signals can be recovered from noise.

12) Calculate and plot the autocorrelation function of a square wave.

13) Draw a neat sketch of an arrangement that could be used to measure the average velocity of a strip of hot steel in a rolling mill.

## Problems on Power Spectral Density

14) Two stationary random signals, $x(t)$ and $y(t)$ are added to produce another random signal $z(t)$.
Determine an expression for the power spectral density of $z(t)$
a)    when $x(t)$ and $y(t)$ are correlated, and
b)    when $x(t)$ and $y(t)$ are uncorrelated.

15) Band limited white noise has a spectral density of $10^{-6}\,V^2$ per Hz in the range $-100kHz.$ to $+100kHz.$ Calculate the RMS value of the noise.

16) A random signal has a power spectral density of

$$S_{XX}(\omega) = \frac{A}{1 + \left(\dfrac{\omega}{\omega_0}\right)^2}.$$

Find the fraction of the total power contained in the frequency range $-\omega_0 \le \omega \le +\omega_0$.

17) White noise of spectral density $A$ volts$^2$ per Hz is passed through an ideal band-pass filter with cut off frequencies $f_1$ and $f_2$. What is the mean square of the signal?

18) $x(t)$ and $y(t)$ are related by the following differential equation:

$$\tau \frac{dy(t)}{dt} + y(t) = x(t).$$

a)    Show that $S_{YY}(\omega) = \dfrac{S_{XX}(\omega)}{1 + \omega^2\tau^2}$.

b)    If $S_{XX}(\omega)$ is a constant $A$, show that the mean square of the signal $y(t)$ is $\dfrac{A}{2\tau}$.

19) A physical process is modelled as a series of rectangular pulses each of duration $\tau$ arriving at random epoch. Derive an expression for the power spectral density function associated with the signal.

20) $x(t)$ and $y(t)$ are related as shown in the following differential equation:

$$\frac{d^2 y(t)}{dt} + 2\xi\omega_n \frac{dy(t)}{dt} + \omega_n^2 y(t) = \omega_n^2 x(t)$$

Show that the power spectral density associated with $y(t)$ is

$$S_{YY}(\omega) = \frac{\omega_n^4 S_{xx}(\omega)}{(\omega^2 - \omega_n^2)^2 + 4\xi^2 \omega_n^2 \omega^2}$$

## Problems on the Wiener-Khinchin Equations.

21) White noise is passed through an ideal band pass filter with cut off frequencies $f_1$ and $f_2$. Derive an expression for the autocorrelation function for the following cases:

a)  $f_1 = 0$ i.e. an ideal **low pass** filter.

b)  $f_2 = f_1 + \delta f_1$, i.e. a **narrow band pass** filter, and

c)  Sketch the autocorrelation function for the first case when the upper frequency cut off tends to infinity.

22) A telegraph signal switches between $+A$ and $-A$ under the governance of the following Poisson distribution:

$$P(k) = \frac{[\mu\lambda]^k}{k!} e^{-\mu\lambda},$$

where $P(k)$ is the probability of there being $k$ changes of sign in the time $\lambda$ and $\mu$ is the average number of zero crossings per unit time. Derive an expression for the power spectral density of the signal.

23) A random signal has a power spectral density given by:

$$S_{xx}(\omega) = \frac{A}{1 + \left(\dfrac{\omega}{\omega_0}\right)^2}.$$

Derive an expression for the autocorrelation function of the signal.

24) $x(t)$ and $y(t)$ are related as shown in the following differential equation:

$$\tau \frac{dy(t)}{dt} + y(t) = x(t).$$

Show that the autocorrelation function is:

$$R_{yy}(\lambda) = \frac{A e^{-\frac{\lambda}{\tau}}}{2\tau}.$$

25) Derive an expression for the power spectral density of a random binary signal.

## Problems of a General Nature

26) White noise of spectral density $A^2$ per Hz is fed to an inductance and a resistor connected in series. An output voltage is taken across the resistor.

- Derive a stochastic equation relating the input and output.
- Calculate the power spectral density of the output.
- Calculate the total mean square of the output.
- Calculate the autocorrelation of the output.
- Calculate the cross-spectral density between output and input.

27) $x(t)$ and $y(t)$ are related as shown in the following differential equation:

$$\frac{d^2 y(t)}{dt} + 2\xi\omega_n \frac{dy(t)}{dt} + \omega_n^2 y(t) = \omega_n^2 x(t)$$

Show that the power spectral density associated with $y(t)$ is

$$S_{YY}(\omega) = \frac{\omega_n^4 S_{xx}(\omega)}{(\omega^2 - \omega_n^2)^2 + 4\xi^2\omega_n^2\omega^2},$$

and from this calculate the autocorrelation function.

28) The autocorrelation function for a stationary random function is given by the expression

$$R_{xx}(\lambda) = A^2 e^{-(\omega_1\lambda)} \cos(\omega_2\lambda)$$

Calculate and sketch the corresponding spectral density function.

## SOLUTIONS TO SELECTED TUTORIAL PROBLEMS

In this section solutions to some of the tutorial problems given in the previous section are presented, although in the interests of brevity some solutions are given in outline form only. The question numbers correspond to those used in the previous section

1)   The solution to this problem lies in using the definitions for mean and mean square given in terms of pdf's, i.e.:

$$\text{mean} = \int_{-\infty}^{+\infty} Xf(X)dX,$$

$$\text{mean square} = \int_{-\infty}^{+\infty} X^2 f(X)dX,$$

and of course that the variance equals the mean square - (the mean$^2$). Substituting the values given, we have:

$$\text{mean} = \int_0^1 X*1*dX = \frac{X^2}{2}\Big|_0^1 = \frac{1}{2}$$

$$\text{mean square} = \int_0^1 X^2 .1dX = \frac{X^3}{3}\Big|_0^1 = \frac{1}{3}$$

and the variance is $1/3 - 1/4 = 1/12$.

5) Problems of this type are based on the knowledge that the area under any pdf is unity. In this case because the pdf is symmetrical we can use the fact that the area under the right-hand plane is precisely one-half, as shown below:

$$\int_0^\infty ae^{-2bX}\,dX = 0.5$$

which, when integrated, yields:

$$\left.\frac{ae^{-2bX}}{-2b}\right|_0^\infty = 0.5$$

from which $a=b$.

6) This question uses exactly the same method as the one above but needs the evaluation of the integral of $\exp(-X^2)$ which can be evaluated exactly when the limits are zero to infinity as shown below:

$$\text{let integral} = \int_0^\infty e^{X^2}\,dX$$

$$\text{and integral} = \int_0^\infty e^{Y^2}\,dY$$

$$\text{hence integral}^2 = \iint e^{-(X^2+Y^2)}\,dX\,dY$$

By changing to polar coordinates this integral can be evaluated, and then the square root taken to recover the value of the integral required.

10) $R_{xx}(\lambda) = A^2\left[\text{sum of probs for an even } k\right] - A^2\left[\text{sum of probs for an odd } k\right]$

$$R_{xx}(\lambda) = A^2\left[p(0) + p(2) + p(4)+...\right] - A^2\left[p(1) + p(3) + p(5)+...\right]$$

$$R_{xx}(\lambda) = A^2\left[e^{-\mu\lambda} + \frac{(\mu\lambda)^2}{2!}e^{-\mu\lambda} + \frac{(\mu\lambda)^4}{4!}e^{-\mu\lambda}+..\right] - A^2\left[\mu\lambda e^{-\mu\lambda} + \frac{(\mu\lambda)^3}{3!}e^{-\mu\lambda}+..\right]$$

Now, by collecting terms from each bracket in ascending order we have:

$$R_{xx}(\lambda) = A^2 e^{-\mu\lambda}\left[1 - \mu\lambda + \frac{(\mu\lambda)^2}{2!} - \frac{(\mu\lambda)^3}{3!} + \frac{(\mu\lambda)^4}{4!}...\right]$$

$R_{xx}(\lambda) = A^2 e^{-2\mu\lambda}$ or, more formally, $A^2 e^{-2\mu|\lambda|}$

11) This example is included to show the effect of a periodic signal buried in a random telegraph signal. The autocorrelation function of the former is periodic, whereas the autocorrelation of the random telegraph signal decreases exponentially as indicated in the solution to 5 above. As the autocorrelation function of two uncorrelated signals is simply the addition of the respective autocorrelation functions we may examine the autocorrelation function of the composite signal at long delays to discover the existence and period of any periodic signal. This is illustrated in the following Figure 2.13

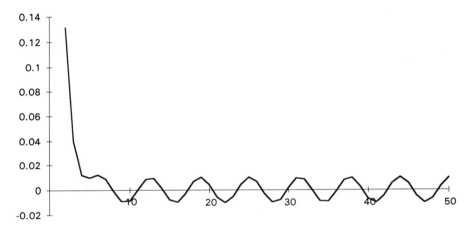

**Figure 2.13** Autocorrelation of a sine wave mixed with a random telegraph wave

15) To avoid the trap in problems of this type, watch out for the range of the frequency scale and whether or not the frequencies are expressed in rad/s or Hz. Perhaps the most appropriate way to proceed is to recognise that the mean square of any signal is expressed in units squared. In the present case the frequency band is 200khz wide (−100khz to +100khz) and the units of spectral density are volts squared per Hz. The total mean square is therefore given as

$$200*10^3*10^{-6}=0.2 \text{ V}^2$$

and the RMS is the square root of this quantity.

18) With this problem we take the Fourier transform of each side of the differential equation to give:

$$[1+ j\omega\tau\}Y(\omega) = X(\omega)$$

and write down the conjugate complex of the above as:

$$[1- j\omega\tau\}Y^*(\omega) = X^*(\omega)$$

Multiplying these together and remembering the definition of spectral density produces

$$S_{YY}(\omega) = \frac{S_{XX}(\omega)}{1 + \omega^2 \tau^2}.$$

The mean square of the signal is the area under the above (allowing for the use of the appropriate units). We can evaluate this expression when the input spectral density is given as a constant A; for example, by direct integration, as follows;

$$\text{mean square} = \frac{1}{2\pi} \int\limits_{-\infty}^{\infty} \frac{A}{1 + \omega^2 \tau^2} d\omega$$

$$\text{mean square} = \frac{A}{2\tau}.$$

20) In this problem the autocorrelation function determined in relation to (10) is used and the Wiener-Khinchin equations applied as follows:

$$S_{XX}(\omega) = \int\limits_{-\infty}^{+\infty} R_{xx}(\lambda) e^{-j\omega\lambda} d\lambda$$

$$S_{XX}(\omega) = \int\limits_{-\infty}^{0} R_{xx}(\lambda) e^{-j\omega\lambda} d\lambda + \int\limits_{0}^{+\infty} R_{xx}(\lambda) e^{-j\omega\lambda} d\lambda$$

$$S_{XX}(\omega) = \int\limits_{-\infty}^{0} A^2 e^{2\mu\lambda} e^{-j\omega\lambda} d\lambda + \int\limits_{0}^{+\infty} A^2 e^{-2\mu\lambda} e^{-j\omega\lambda} d\lambda$$

$$S_{XX}(\omega) = A^2 \left[ \frac{e^{(2\mu - j\omega)\lambda}}{2\mu - j\omega} \right]_{-\infty}^{0} + A^2 \left[ \frac{e^{-(2\mu + j\omega)\lambda}}{-(2\mu + j\omega)} \right]_{o}^{+\infty}$$

$$S_{XX}(\omega) = A^2 \cdot \left[ \frac{1}{2\mu - j\omega} + \frac{1}{2\mu + j\omega} \right]$$

22) Taking Fourier transforms of both sides of the differential equation gives

$$(j\omega\tau + 1).Y(\omega) = X(\omega),$$

and then taking the complex conjugate of the above gives

$$(-j\omega\tau + 1).Y^*(\omega) = X^*(\omega)$$

where the * denotes the complex conjugate. Multiplying these together and remembering the definition of spectral density produces

$$Y(\omega).Y^*(\omega) = \frac{X(\omega).X^*(\omega)}{1+\omega^2\tau^2}.$$

Now, remembering the definition of power spectral density we take averages of both sides of the above to yield the desired result of

$$S_{YY}(\omega) = \frac{S_{xx}(\omega)}{1+\omega^2\tau^2}$$

$$R_{yy}(\lambda) = \frac{1}{2\pi}\int_{-\infty}^{\infty}\frac{Ae^{j\omega\lambda}d\omega}{1+\omega^2\tau^2}$$

$$R_{yy}(\lambda) = \frac{1}{2\pi\tau^2}\int_{-\infty}^{\infty}\frac{Ae^{j\omega\lambda}d\omega}{(1+j/\tau).(1-j/\tau)}.$$

This integrand has two poles at $\pm j/\tau$; therefore, substituting for the one in the upper half-plane gives

$$R_{yy}(\lambda) = \frac{1}{2\pi\tau^2}.2\pi j\frac{Ae^{-\lambda/\tau}}{2j/\tau}$$

$$R_{yy}(\lambda) = \frac{Ae^{-\lambda/\tau}}{2\tau}.$$

27) The solution to this problem is given as part of the explanation of the practical data exercises on a randomly vibrating cantilever discussed in section 8.3.

# 3

# SAMPLED DATA PROCESSES AND SIGNAL CONDITIONING

## 3.1    INTRODUCTION AND LEARNING OBJECTIVES

In practice, the continuous signals monitored on a plant are usually sampled at equal intervals of time to produce a time series record of the process behaviour. In order to confidently extract information from a time series record, it is necessary to employ the mathematics of discrete processes, which are reviewed in this Chapter.

The topics covered in this Chapter include:

- continuous and discrete processes and how they are derived from data sampled in the time domain;
- Z-transforms, sampled data processes and discrete process models;
- discrete Fourier transforms and the frequency response of linear processes;
- the mathematical steps for deriving equivalent discrete process models from their continuous process models given by ordinary differential equations (ODEs);
- the application of discrete process models and Z-transforms to obtain the frequency response of the discrete process;
- comparison of the time and frequency response characteristics of continuous and discrete models;

## 3.2    SAMPLED SIGNALS AND Z-TRANSFORMS

### 3.2.1    Continuous and Discrete Processes

The signals monitored on a process are classed as being either *continuous* or *discrete* in time. *Continuous* signals have values at *any* point in a given range, such that any range or sub-range constitutes an *infinite* set of values. *Discrete* signals, on the other hand, may take on only a *finite* set of values in any given range or sub-range. The difference between the two types of signals is depicted in Figure 3.1. In the present context, the discrete values are equally spaced within a range, but in a more general context the spacing may vary considerably

**Figure 3.1** Comparison of continuous and discrete quantities

to satisfy numerical resolution requirements (e.g. space-time numerical integration methods).

It follows that a *continuous process* is one which is characterised by, or identified from, continuous signals, and a *discrete process* is one which is characterised by, or identified from, discrete signals. Since computers are digital machines, discrete processes have become an increasingly important field of study in the latter half of this century.

### 3.2.2    Sampling Theorems, Aliasing and Signal Reconstruction

Sampled signals are obtained from an analogue-to-digital (A/D) coverter or from inherently sampled data such as daily, monthly or annual statistics relating to commercial or meteorological activities. In the former case, the continuous (or 'analogue') input and output signals $x(t)$ and $y(t)$ of a linear process (termed analogue to analog or $A \rightarrow A$, Figure 3.2) are sampled at time intervals $\Delta$ apart such that corresponding data sequences (or time series) are obtained. These sampled sequences, or *impulse-modulated signals*, can be written as:

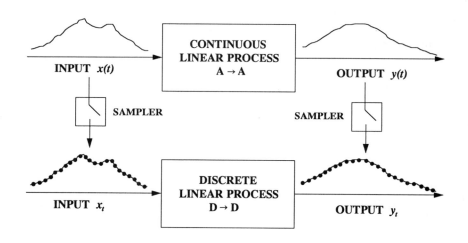

**Figure 3.2** Continuous and discrete linear processes

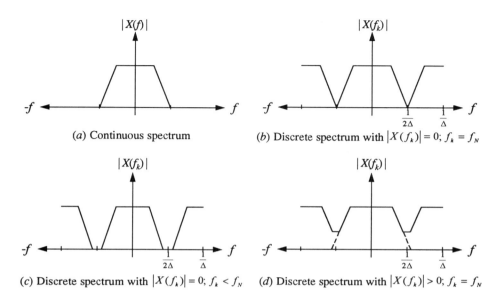

(a) Continuous spectrum

(b) Discrete spectrum with $|X(f_k)| = 0$; $f_k = f_N$

(c) Discrete spectrum with $|X(f_k)| = 0$; $f_k < f_N$

(d) Discrete spectrum with $|X(f_k)| > 0$; $f_k = f_N$

**Figure 3.3** Transforms of continuous and discrete signals for various sampling intervals

$$f_t = f(t)\sum_{k=-\infty}^{\infty}\delta(t - k\Delta) = \sum_{k=-\infty}^{\infty}f(k\Delta)\delta(t - k\Delta) \qquad \dots (3.1)$$

where      $f(t)$ is an arbitrary function, e.g. $x(t)$ or $y(t)$,

     $k$ denotes the $k$th time step, and

     $\delta(t-k\Delta)$ is the Kronecker delta function which has the property

     $\delta(t-k\Delta)$   $= 1$    if $t = k\Delta$

               $= 0$    if $t \neq k\Delta$

Thus, the input and output data sequences are given by:

$$x_t = \sum_{k=-\infty}^{\infty}x(k\Delta)\delta(t - k\Delta); \quad y_t = \sum_{k=-\infty}^{\infty}y(k\Delta)\delta(t - k\Delta) \qquad \dots (3.2)$$

Equations (3.2) apply to a digital sampler with an infitesimal pulse width $\Delta_p$. The Fourier transform of a rectangular pulse of unit amplitude and width $2\Delta_p$ is the *sinc function*:

$$\delta(\omega) = 2\Delta_p\frac{\sin(\omega\Delta_p)}{\omega\Delta_p} = 2\Delta_p\,\text{sinc}(\omega\Delta_p) \qquad \dots (3.3)$$

This sinc function is the ideal filter for recovering the continuous signal $x(t)$ from the sampled signal $x(k\Delta)$ or, alternatively, it is the ideal interpolating function for equally spaced ordinates. The Fourier transforms of a continuous signal $x(t)$ and its sampled signal $x(k\Delta)$ are shown in Figure 3.3. If the sampling interval ($\Delta$) is such that $|X(f_k)|$ falls to zero on or before $|f| = 1/(2\Delta)$, as in Figures 3.3($b$) and 3.3($c$), then it is possible to recover $x(t)$ from the sampled signal. However, if $|X(f_k)|$ is *not* zero at $|f| = 1/(2\Delta)$, then components above this frequency will appear in the sample spectrum, as shown in Figure 3.3($d$). The frequency

$$f_N = \frac{1}{2\Delta} \qquad \qquad \text{... (3.4)}$$

is the highest frequency which can be detected when data are sampled at time intervals $\Delta$ apart, and is known as the **Nyquist frequency**.

*Aliasing*

If the signal being sampled has frequency components greater than the sampling frequency, then a phenomenon known as **aliasing** occurs, Figure 3.3($d$). This phenomenon is depicted graphically in the time domain in Figure 3.4, which shows a 'high' frequency signal being sampled by a lower frequency sampler. The resulting digitised values, given by the black dots at the intersection of the two sinusoids, bear little resemblance to the true signal. Two practical examples of aliasing are:

- the *strobing effect* of a movie camera shutter as it films the slow forward or backward rotation of the wheel spokes of an moving wagon in a western movie, and

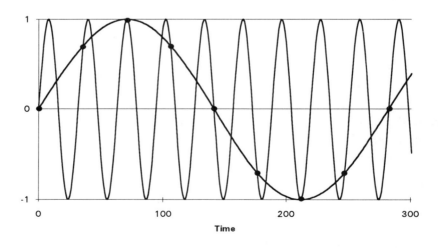

**Figure 3.4** Graphical representation of aliasing

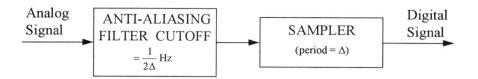

**Figure 3.5** Procedure for digitising an analog signal

- the slow forward or backward rotation of an engine flywheel marker near the *strobe frequency* of an engine timing light.

In each case, the 'speed' of the slow rotation is given by the difference between the rotational frequency $(f_r)$ and the sampling frequency $(f_N)$, and will be *forward* if the rotation frequency is *greater* than the sampling frequency (i.e. $f_r > f_N$), and it will be *backwards* if the rotation frequency is *less* than the sampling requency (i.e. $f_r < f_N$).

Modern digital signal processing (DSP) equipment usually incorporate *anti-aliasing* filters which low-pass filter the analogue signal at a cutoff frequency slightly less than the sampling frequency $(<\frac{1}{2}\Delta$ Hz) *before* being passed through the sampler (Figure 3.5).

### Signal Reconstruction

There is a loss of information (energy) in sampling continuous signals (Figure 3.6($a$)), and **signal reconstruction** techniques are often used to reconstruct the original signal by a process called *demodulation*, in which *data extrapolators* or *data holds* are employed. The simplest data hold is the *zero-order* data hold (Figure 3.6($b$)) in which the demodulated signal at any instant in time is equal to the last-received sample. In some applications, a zero-order hold may yield only crude approximations of the original signal, and so *first-order* and *higher order* holds using nonlinear functions are sometimes employed. We will look at data holds again when considering the relationship between discrete and continuous processes. Note that, in utilising data holds, the *pulse width* $(\Delta_p)$ is equal to the *sample interval* $(\Delta)$.

### 3.2.3 The Z-Transform

### Definition

The analysis of *discrete* or *sampled data* processes from functional relationships between strings of numbers (samples) requires a *shift operation* back and forth in time. For an arbitrary continuous function $f(t)$ sampled at intervals $\Delta$ apart such that the sequence $f_i$ is obtained (see equations (3.2)):

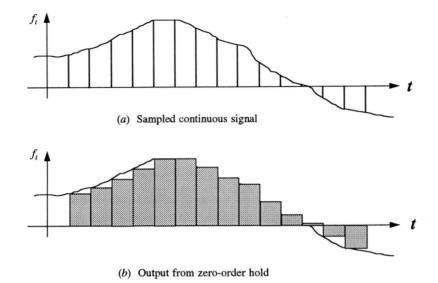

(a) Sampled continuous signal

(b) Output from zero-order hold

**Figure 3.6** Signal reconstruction with zero-order hold

$$f_t = \sum_{k=-\infty}^{\infty} f(k\Delta)\delta(t - k\Delta) \qquad \qquad ... (3.5)$$

Taking the Laplace transform yields:

$$F_s(s) = \int_0^\infty \sum_{k=0}^\infty f(k\Delta)\delta(t - k\Delta)e^{-st}dt$$

Interchanging the integration and summation yields

$$F_s(s) = \sum_{k=0}^\infty f(k\Delta)\int_0^\infty \delta(t - k\Delta)e^{-st}\,dt = \sum_{k=0}^\infty f(k\Delta)e^{-sk\Delta} \qquad ... (3.6)$$

Defining the complex variable $z$ as

$$z = e^{s\Delta} \qquad \qquad ... (3.7)$$

yields the one-sided $Z$-transform definition of the sequence of samples $f_t \equiv f(k\Delta)$, $t>0$:

$$F(z) = Z\{f_t\} = f_0 + f_1 z^{-1} + f_2 z^{-2} + ... + f_k z^{-k} + ... = \sum_{t=0}^\infty f_t z^{-t} \qquad ... (3.8)$$

*Z-Transform Properties*

Like the Fourier transform, which will be discussed later, the Z-transform has a number of properties which are tabulated in Appendix A3. Of these, one of the most important is the *time shift* property (2):

$$f_{t-d} = z^{-d} f_t \quad \rightarrow \quad f_{t-1} = z^{-1} f_t$$
$$f_{t+d} = z^{d} f_t \quad \rightarrow \quad f_{t+1} = z f \qquad \qquad \dots (3.9)$$

where $d$ denotes time delay (−ve) or $d$-step ahead forecast (+ve).

*Examples*

Three examples are presented to demonstrate the use of Z-transforms. In two of the proofs that follow, note the use of the power series relationship:

$$1 + r + r^2 + r^3 + r^4 + \dots = \sum_{k=0}^{\infty} r^k = \frac{1}{1-r} \qquad \qquad \dots (3.9)$$

1) *Discrete step function* given by $f_n = 1, 0 \le n \le \infty$, then

$$F(z) = Z\{f_n\}$$
$$= f_0 + f_1 z^{-1} + f_2 z^{-2} + \dots = 1 + z^{-1} + z^{-2} + \dots \qquad \dots (3.10a)$$
$$= \frac{1}{1 - z^{-1}} = \frac{z}{z - 1}$$

2) *Decaying exponential* given by $f_n = e^{-an\Delta}, 0 \le n \le \infty$, then

$$F(z) = Z\{f_n\} = f_0 + f_1 z^{-1} + f_2 z^{-2} + \dots$$
$$= 1 + e^{-a\Delta} z^{-1} + e^{-2a\Delta} z^{-2} + \dots \qquad \qquad \dots (3.10b)$$
$$= \frac{1}{1 - e^{-a\Delta} z^{-1}} = \frac{z}{z - e^{-a\Delta}}$$

3) *Unit Pulse* given by $f_n = 1, n = 0; f_n = 0, n > 0$, then

$$F(z) = Z\{f_n\} = 1 + 0z^{-1} + 0z^{-2} + \dots = 1 \qquad \qquad \dots (3.10c)$$

### 3.2.4 Difference Equations and Discrete Process Models

The above review of the theory on Z-transforms provides a basis for developing the mathematical relationship between continuous and discrete process models. In general, a continuous process may be represented by the linear ordinary differential equation:

$$\sum_{j=0}^{n} a_j(t) \frac{d^j y(t)}{dt^j} = \sum_{k=0}^{m} b_k(t) \frac{d^k x(t)}{dt^k} \qquad \text{... (3.11)}$$

where   $a_j(t)$, $b_k(t)$   are the time variant/invariant coefficients/parameters
        $x(t)$            is the process input or forcing function, and
        $y(t)$            is the process output or response.
For a pulsed input (i.e. no data hold operation), this process is coincident
with the discrete process satisfying the difference equation

$$\sum_{j=0}^{n} \alpha_j y_{t-j} = \sum_{k=0}^{m} \beta_k x_{t-k} \qquad \text{... (3.12)}$$

This discrete model is often referred to as an **autoregressive-moving average** ARMA($n,m$) model in which the $\alpha$'s are the **autoregressive** parameters and the $\beta$'s are the **moving average** parameters, respectively.

*Case 1: Data Smoothing*

Suppose we have a string of numerical samples ($x_n$ say) of a signal (or points on a graph) and the signal is *noisy*. An obvious way to 'smooth' the data is to use the running average:

$$y_n = \frac{1}{3}[x_n + x_{n-1} + x_{n-2}] \qquad \text{... (3.13)}$$

Three (3) samples are chosen here by way of example only. More efficient, higher order data smoothing algorithms, or *low-pass digital filters*, are normally used in practice.

### Difference Operator Form

Before concluding this section on difference equations, we should note that the differential equation (3.11) is often expressed as a linear difference equation:

$$\sum_{j=0}^{n} \gamma_j \delta^j = \sum_{k=0}^{m} \theta_k \delta^k \qquad \text{... (3.14)}$$

where the *difference operator* $\delta$ has several forms depending on the numerical difference algorithm, e.g. for a backward difference trapezoidal scheme $\delta x_t = (x_t - x_{t-1}) / \Delta$. It follows that, since there is a linear relationship between the $\delta$-operator and the Z-transform operator (e.g. $z^{-1} = 1 - \delta\Delta$ for the trapezoidal case), the above difference equations using the $z$-operator can be written in terms of the $\delta$-operator to yield equivalence relations between the corresponding parameters. More importantly, the properties of the Z-transform operator also apply to the δ-operator.

*Case 2: Second Order Discrete System*

From equation (3.11), a second order differential equation for a continuous process is given by:

$$a_2 \frac{d^2 y(t)}{dt^2} + a_1 \frac{dy(t)}{dt} + a_0 y(t) = b_0 x(t) \qquad \dots (3.15)$$

Equation (3.15) can be approximated by the difference equation (with $d/dt \equiv \delta$)

$$a_2 \delta^2 y_t + a_1 \delta y_t + a_0 y_t = b_0 x_t \qquad \dots (3.16)$$

and hence, with $\delta x_t = (x_t - x_{t-1}) / \Delta$, we obtain

$$(a_2 + a_1\Delta + a_0\Delta^2) y_t - (2a_2 + a_1\Delta) y_{t-1} + a_2 y_{t-2} = b_0\Delta^2 x_t \qquad \dots (3.17)$$

from which it follows that, using equation (3.12),

$$\alpha_0 = a_2 + a_1\Delta + a_0\Delta^2; \quad \alpha_1 = -(2a_2 + a_1\Delta); \quad \alpha_2 = a_2$$
$$\beta_0 = b_0\Delta^2 \qquad \dots (3.18)$$

Note that this approximation differs from that obtained using the (zero-order data hold) equivalence relationships between continuous and discrete processes derived in Appendix B3. In general, discrete transfer function models are of importance in their own right and do not need to be justified in terms of their equivalent continuous transfer function models. However, there are applications where such relationships are of interest.

### 3.2.5 Process Transfer Functions

The transfer function $G(s)$ of a continuous process is derived from equation (3.11) by taking the Laplace transform and rearranging to yield (assuming zero initial conditions):

$$G(s) = \frac{Y(s)}{X(s)} = \frac{B(s)}{A(s)} = \frac{\sum\limits_{k=0}^{m} b_k s^j}{\sum\limits_{j=0}^{n} a_j s^k} \qquad \dots (3.19)$$

where $X(s)$ is the Laplace transform of the process input $x(t)$,

$Y(s)$ is the Laplace transform of the process output $y(t)$,

$A(s)$ is a polynomial of transfer function zeros, and

$B(s)$ is a polynomial of transfer function poles.

In a similar manner, equation (3.19) can be written in Z-transform notation as:

$$\left(\sum_{j=0}^{n}\alpha_j z^{-j}\right)y_t = \left(\sum_{k=0}^{m}\beta_k z^{-k}\right)x_t \qquad \qquad \text{... (3.20)}$$

and by letting

$$A(z) = \alpha_0 + \alpha_1 z^{-1} + ... + \alpha_n z^{-n} = \sum_{j=0}^{n}\alpha_j z^{-j}$$

$$\text{... (3.21)}$$

$$B(z) = \beta_0 + \beta_1 z^{-1} + ... + \beta_m z^{-m} = \sum_{k=0}^{m}\beta_k z^{-k}$$

then equation (3.20) can be written as:

$$A(z)\,y_t = B(z)\,x_t \qquad \qquad \text{(3.22)}$$

that is, the **discrete process transfer function** $G(z^{-1})$ is given by

$$G(z) = \frac{Y(z)}{X(z)} = \frac{B(z)}{A(z)} \qquad \qquad \text{... (3.23)}$$

The discrete process transfer function $G(z)$, equation (3.21), is depicted in Figure 3.7 above. In expanded form, equation (3.22) is given by

$$y_t = g_0 x_t + g_1 x_{t-1} + ... + g_N x_{t-N} = \sum_{l=0}^{N}g_l x_{t-l}$$

$$\text{... (3.24)}$$

$$G(z) = g_0 + g_1 z^{-1} + ... + g_N z^{-N} = \sum_{l=0}^{N}g_l z^{-l}$$

where $N$ is the number of data samples. In practice, the series is usually truncated to a lesser number terms because the remaining terms have negligible

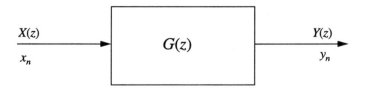

**Figure 3.7** Discrete process transfer function

affect on the overall result. Note that equation (3.24) is now an $N$th order **moving average** (MA) model.

Using the definitions for $A(z)$ and $B(z)$, equation (3.21), and $G(z)$, equation (3.24), the equating the coefficients in the powers of $z^{-1}$ of the polynomial product $B(z) = A(z)G(z)$ yields the following relationship between the parameters of the transfer function $G(z)$ and its rational functions, $A(z)$ and $B(z)$, as:

$$g_k = [\beta_k - \sum_{j=1}^{n} g_{k-j}\alpha_j]/\alpha_0 \qquad \text{... (3.25)}$$

where $\beta_k = 0$ $(k > m)$ and $g_{k-j} = 0$ $(k < j)$. Note that $\alpha_0 = 1$ without loss of generality, and $\beta_0 = 0$ for zero initial conditions.

### Process Time (Transport) Delays

In many processes, it takes a finite time for material to flow from the input to the output of a process, and this *time delay* (or *transport delay*) can be significant in large processes. A simple example is the time delay for water to reach the end of an irrigation channel or garden hose after the weir gate or garden tap has been opened/turned on.

For discrete processes with a time delay of $d$ time steps, equation (3.12) becomes:

$$\sum_{j=0}^{n} \alpha_j y_{t-j} = \sum_{k=0}^{m} \beta_k x_{t-k-d} \qquad \text{... (3.26)}$$

where the output $y_t = 0$ for $t < d$ (physical realisability condition). In Z-transform notation, equation (3.22) becomes

$$y_t = G(z)x_t = \frac{B(z)z^{-d}}{A(z)}x_t \qquad \text{... (3.27)}$$

where the time shifting property of Z-transforms has been invoked. In a similar manner to equation (3.25), the corresponding time delay relationship between the parameters of $G(z)$ and its rational functions $A(z)$ and $B(z)$ is given by

$$g_k = [\beta_{k-d} - \sum_{j=1}^{n} g_{k-j}\alpha_j]/\alpha_0 \qquad \text{... (3.28)}$$

where $\beta_{l-d} = 0$ $(m < l - d < 0)$ and $g_{l-k} = 0$ $(l < k)$.

*Conversion to the Frequency Domain*

It is often useful to compare continuous and discrete process transfer function models in the frequency domain. This is achieved by noting that, for a finite pulse of duration $\Delta$, the Z-transform operator expressed as a function of frequency ($f$) is given by (see equation (3.7)):

$$z = e^{s\Delta} = e^{j\omega\Delta} = \cos(\omega\Delta) + j\sin(\omega\Delta)$$

$$s = j\omega \quad (\sigma = 0); \quad \omega = 2\pi f$$

... (3.29)

where $s$ is the Laplace operator and $\omega$ is the angular frequency (radians/s). Thus the discrete frequency response

$$G(j\omega) = G(z)\big|_{z^{-1} = e^{-j\omega\Delta}}$$

... (3.30)

## 3.3    DISCRETE FOURIER TRANSFORMS

### 3.3.1    The Discrete Fourier Transform (DFT)

In Chapter 2, section 2.4, we reviewed the concept of Fourier transforms in some detail because it forms the theoretical basis for the numerical computation of signal spectra from discrete (or sampled) data. This is done using the *Discrete Fourier Transform* (DFT).

Recall from equation (3.2) that a continuous signal $x(t)$ sampled at a time interval $\Delta$ apart yields (in principle) the *infinite* discrete time series of samples given by

$$x_t = x(t) \sum_{k=-\infty}^{\infty} \delta(t - k\Delta) = \sum_{k=-\infty}^{\infty} x(k\Delta)\delta(t - k\Delta); \quad t \to \infty \qquad ... (3.31)$$

To convert this *infinite* series to a *periodic* series over $N$ samples, Figure 3.8, we *convolve* it with a *railing function* $r(t)$ given by

$$r(t) = T \sum_{m=-\infty}^{\infty} \delta(t - mT) \qquad ... (3.32)$$

where the scaling by by the period $T$ cancels the effects of Fourier transforming the railings function:

$$FT\left[ \sum_{m=-\infty}^{\infty} \delta(t - mT) \right] = \frac{1}{T} \sum_{m=-\infty}^{\infty} \delta\left( f - \frac{m}{T} \right) \qquad ... (3.33)$$

Performing the convolution operation to make the time samples periodic, yields:

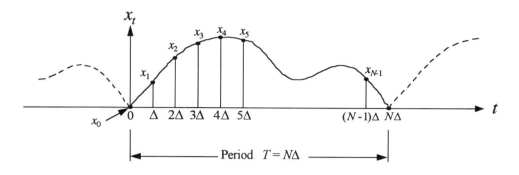

**Figure 3.8** Periodic discrete time data samples

$$x(t) = x_T(t) * r(t) = \sum_{n=0}^{N} x(n\Delta)\delta(t - n\Delta) * T \sum_{m=-\infty}^{\infty} \delta(t - mT) \qquad \dots (3.34)$$

Taking the Fourier Transform of this expression (noting that convolution→ product) yields:

$$\widetilde{X}(f) = \left[ \sum_{n=0}^{N-1} x(n\Delta)e^{-j2\pi f n\Delta} \right] \cdot \left[ \sum_{m=-\infty}^{\infty} \delta\left( f - \frac{m}{T} \right) \right] \qquad \dots (3.35)$$

This is a railing function in frequency modulated by the sinusoids in the transform of $x(t)$. The function $\widetilde{X}(f)$ only has values at frequencies $f_k = k/T$, and these are repeated after the first $N$ values (Figure 3.8). The first $N$ samples of $\widetilde{X}(f)$ are thus given by the *Discrete Fourier Transform* (DFT) equation:

$$X(f_k) = \sum_{n=0}^{N-1} x(n)e^{-j2\pi k n\Delta/T}; \quad k = 0, N - 1 \qquad \dots (3.36)$$

where the frequency values are a 'mirror image' around the 'folding frequency' or *Nyquist sampling frequency* ($k = N/2$, see section 3.2.1 and Figure 3.9). The *Inverse Discrete Fourier Transform* (IDFT) is derived in a similar fashion as:

$$x(n) = \frac{1}{N} \sum_{k=0}^{N-1} X(f_k)e^{j2\pi k n\Delta/T}; \quad n = 0, N - 1 \qquad \dots (3.37)$$

As with the continuous Fourier Transform, equations (3.36) and (3.37) form a 'DFT pair'.

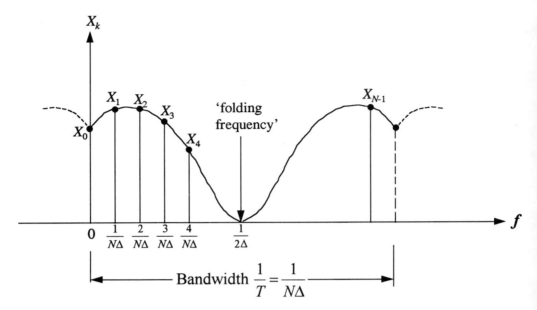

**Figure 3.9** Fourier transform of periodic discrete time series

### 3.3.2    The Fast Fourier Transform (FFT)

A drawback in utilising the DFT equation (3.36) for discrete signal processing, is the time it takes to compute the transform values at all frequencies for large arrays of discrete data ($N$ very large). More efficient numerical algorithms were developed to improve computational speed, and this has resulted in numerous *Fast Fourier Transform* (FFT) algorithms being published over the years. The underlying concept of a typical FFT algorithm is outlined below, but it should be noted that it is only one of several ways of reducing the computational effort for a DFT. Note also that the time function, $x(n\Delta)$, can also be *complex* if required.

Consider the DFT equation (3.35):

$$X(f_k) \equiv X_k = \sum_{n=0}^{N-1} x(n)e^{-j2\pi kn\Delta/T}; \quad -(N-1) \le k \le N-1$$

Let the exponential term in this equation be given by:

$$W_N = e^{-j2\pi/N} \qquad \qquad \text{... (3.38)}$$

then the summation term in the above equation looks like

$$X_0 = x_0 W_N^0 + x_1 W_N^0 + x_2 W_N^0 + ... + x_{N-1} W_N^0$$

$$X_1 = x_0 W_N^0 + x_1 W_N^1 + x_2 W_N^2 + ... + x_{N-1} W_N^{N-1}$$

$$X_2 = x_0 W_N^0 + x_1 W_N^2 + x_2 W_N^4 + ... + x_{N-1} W_N^{2N-2}$$

$$\bullet \; = \qquad\qquad \bullet \qquad\qquad \bullet$$

$$\bullet \; = \qquad\qquad \bullet \qquad\qquad \bullet$$

$$X_{N-1} = x_0 W_N^0 + x_1 W_N^{N-1} + x_2 W_N^{2N-2} + ... + x_{N-1} W_N^{(N-1)^2}$$

... (3.39)

It can be seen that this involves a lot of complex multiplications and additions. The complex function $W_N$ keeps revolving around a unit circle as it is raised to increasing integer powers, and this fact is exploited by FFT algorithms. Equation (3.39) shows that the complete computation of the DFT will require $N^2$ complex multiplications and $N(N-1)$ complex additions. For large $N$, this is approximately $N^2$ complex multiplications and additions.

Let $N$ be a *binary* number of data points such that

$$N = 2^M \qquad\qquad ... (3.40)$$

then we can compute $X_k$ by separating $x(n)$ into alternate *odd* and *even* numbered samples:

$$X_k = \sum_{n=0}^{N-1} x_n W_N^{kn} = \sum_{n \text{ odd}} x_n W_N^{kn} + \sum_{n \text{ even}} x_n W_N^{kn}$$

$$= \sum_{m=0}^{(N/2)-1} x_{2m} W_N^{2mk} + \sum_{m=0}^{(N/2)-1} x_{(2m+1)} W_N^{(2m+1)k}$$

... (3.41)

However, since

$$W_N^2 = e^{-j2(2\pi/N)} = e^{-j2\pi/(N/2)} = W_{N/2} \qquad\qquad ... (3.42)$$

equation (3.41) becomes

$$X_k = \sum_{m=0}^{(N/2)-1} x_{2m} W_{N/2}^{mk} + W_{N/2}^k \sum_{m=0}^{(N/2)-1} x_{(2m+1)} W_{N/2}^{mk} \qquad\qquad ... (3.43)$$

Each summation on the RHS is an $(N/2)$-point DFT, the first being the sum of *even*-numbered data points and the second the *odd*-numbered data points of the original data sequence. Although the index $k$ ranges over $N$ values, $k=0,1,..,N-1$, each of the sums is computed for $k = 0$ to $(N/2)-1$ since each is periodic in $k$ with period $N/2$. The number of complex computations required now becomes $2(N/2)^2 + N$, which is less than $N^2$ previously.

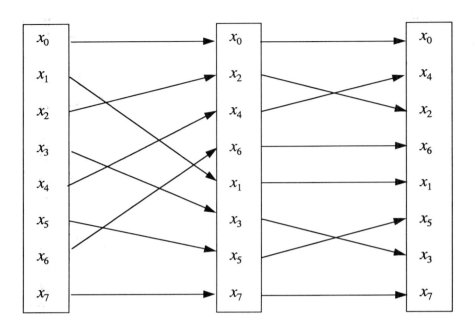

**Figure 3.10** Decimation in time of an 8-point data sequence

This process of splitting the $N$-point DFT into two $(N/2)$-point DFT's may be repeated until, finally, a set of 4-point DFT's is split into $2M$–1 2-point DFT's. At each split alternate samples of the signal are separated out, and an example of this for an 8-point DFT is shown in Figure 3.10. The signal is thus shuffled and sorted at the 'time' end of the FFT process, and is known as *decimation in time*. The end result of this reduction process is that the FFT computation requires $N\log_2(N)$ complex additions and $(N/2)\log_2(N)$ complex multiplications. The savings accrued are indicated in the Table below.

| $N$ | $N^2$ (DFT) | $N \log_2 N$ (FFT) |
|---|---|---|
| 2 | 4 | 2 |
| 8 | 64 | 24 |
| 32 | 1 024 | 160 |
| 128 | 16 384 | 896 |
| 512 | 262 144 | 4 608 |
| 2 048 | 4 194 304 | 22 528 |

## 3.4    SIGNAL CONDITIONING AND PREPROCESSING

A 'raw' signal monitored on a process often requires preprocessing *before* it is subjected to detailed analysis. This section outlines key preprocessing operations designed to ensure reliable results.

### 3.4.1    Removal of the Mean or Average

Recall that a typical signal $x(t)$ normally has a *mean* or *average* value, Figure 2.7, which is defined by:

$$\bar{x}_T = \frac{1}{T}\int_0^T x(t)\,dt \qquad \text{... (3.44)}$$

and, for discrete data sequences, by:

$$\bar{x}_N = \frac{1}{N}\sum_{n=0}^{N} x_n \qquad \text{... (3.45)}$$

The *on-line* computation of the mean is often desired, and this can be accomplished by one of the following procedures:

1) Continuous signals may be filtered by a *low pass filter* with one or more large time constants; this integrates the signal in a similar manner to equation (3.44).
2) Discrete signals may be filtered by a *low pass digital filter* or by the following *moving average* form of equation (3.45) at the $N$th timestep:

$$\bar{x}_N = \frac{1}{N}\sum_{n=0}^{N} x_n = \frac{1}{N}\left[\sum_{n=0}^{N-1} x_n + x_N\right] = \frac{1}{N}[(N-1)\bar{x}_{N-1} + x_N] \quad \text{... (3.46)}$$

which utilises the mean value $\bar{x}_{N-1}$ computed in the previous step. Equation (3.46) tends to have a slow convergence so that initial mean value estimates are in error, and for large $N$ it is insensitive to any non-stationary *drifts* in the mean level. For this reason it should be used with caution. Note, from our previous discussion, that the discrete filter transfer function of equation (3.46) is $G(z^{-1}) = b_0 / (1 - a_1 z^{-1})$, where $b_0 = 1/N$ and $a_1 = (N-1)/N$.

### 3.4.2    Removal of Trends and Drifts

There is a wide variety of non-stationary trends and drifts that can occur in practice, and which may influence downstream functional analysis results. Two common examples are depicted in Figure 3.11. There is no unique method

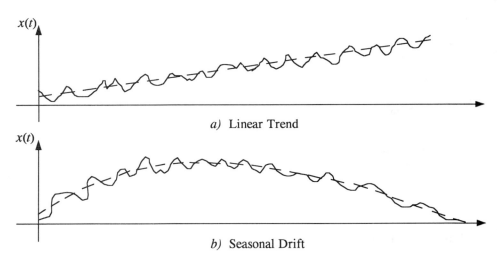

*a)* Linear Trend

*b)* Seasonal Drift

**Figure 3.11** Linear and seasonal drifts

which will remove all types of trends and drifts, but some procedures commonly used in practice include:

1) *Low pass filtering* (analog or digital): the signal is passed through the low pass filter with a low cut off frequency and then subtracted from the 'raw' signal on a *feedforward* or *feedback* basis. This method is useful for removing *linear trends* (either increasing or saw-tooth functions of time) and low frequency *seasonal drifts*. It should not be used if the lower bound of the frequency bandwidth of the signal under investigation is of the same order as the filter cut off frequency.

2) *Adaptive low pass filtering* (digital): this is similar to method (1) except that the digital filter has a parameter (or covariance) resetting facility in the adaptive parameter estimation algorithm for situations where the industrial process moves from its current operating state to a new steady state mode of operation. It is applicable to removing trends and drifts as well as (controlled) step changes in the steady state set point of operation.

3) *Bandpass filtering*: this is perhaps the most commonly used method, but it is only recommended if the lower bound of the bandpass filter bandwidth for the signal under investigation is significantly higher than the low frequency cut-off needed to remove the trend or drift.

4) *High pass filtering*: this is also a commonly used method but, as with (3), it is only recommended if the lower bound of the high pass filter bandwidth for the signal under investigation is significantly higher than the low frequency cut-off needed to remove the trend or drift.

The main reason for the precautions noted above is that indiscriminate application of a procedure can add unwanted 'noise' to the signal which may be spuriously correlated at low frequencies with other similarly processed signals.

### 3.4.3 Digital Filtering

A digital filter is an algorithm which, like an analogue filter, modifies the spectral characteristics of discrete data. Digital filter design is an extensive field of study in its own right, and interested readers are referred to texts listed at the end of this Chapter for further details. Only a brief overview is given in the following discussion.

Recall, from our previous discussion on discrete processes that they can be represented by a discrete transfer function equation (3.23) of the form (see Figure 3.7):

$$G(z) = \frac{Y(z)}{X(z)} = \frac{B(z)}{A(z)} \qquad \ldots (3.47)$$

Transformation of this equation back into the time domain yields the linear difference equation (cf. equations (3.20) and (3.22)):

$$y_t = -\alpha_1 y_{t-1} - \alpha_2 y_{t-2} - \ldots - \alpha_n y_{t-n} + \beta_0 x_t + \beta_1 x_{t-1} + \ldots + \beta_m x_{t-m} \quad \ldots (3.48)$$

The frequency response of the digital filter is given by:

$$G(\omega) = G(z)\big|_{z^{-1} = e^{-j\omega\Delta}} \qquad \ldots (3.49)$$

where the frequency values are limited to the range $-f_N \leq f \leq f_N$ ($f_N$ = Nyquist frequency) in order to avoid aliasing. As with analog filters, the types of digital filters can include: lowpass, highpass, bandpass and bandstop (or notch) filters, and their respective frequency responses are depicted in Figure 3.12.

Digital filter design can be attempted in one of the following ways:

1) conversion of an analogue filter transfer function $G(s)$ into the discrete domain, or
2) direct application of a suitable digital filter $G(z)$ with the desired frequency response.

The first method involves obtaining the Z- transform of $G(s)$, that is,

$$G(z) = Z\{G(s)\} = g_0 + g_1 z^{-1} + \ldots + g_N z^{-N} = \sum_{l=0}^{N} g_l z^{-l} \qquad \ldots (3.50)$$

(cf. equation (3.24)). Equivalent relationships may be determined from standard

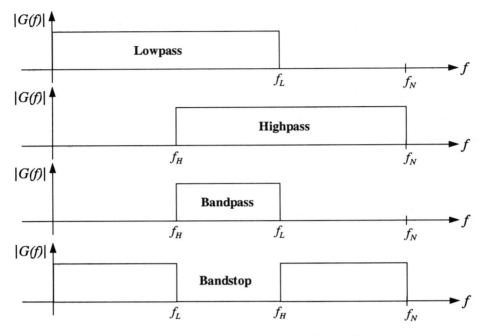

**Figure 3.12** Types of analogue and digital filters

Tables, or if $G(s)$ readily devolves into partial fractions

$$G(s) = \sum_{i=1}^{N} \frac{K_i}{s + a_i} \qquad \qquad \text{... (3.51)}$$

then it can be readily shown that (see Appendix B3)

$$G(z) = \sum_{i=1}^{N} \frac{K_i}{1 - e^{-a_i \Delta} z^{-1}} \qquad \qquad \text{... (3.52)}$$

Alternatively, we can use the *bilinear transform* (BLT) substitution

$$s = \frac{2 \,(1 - z^{-1})}{\Delta \,(1 + z^{-1})} \qquad \qquad \text{... (3.53)}$$

to obtain the equivalent BLT digital filter. Note that the BLT digital filter matches its analogue equivalent more closely as the sample rate *increases*, that is, as $\Delta \to 0$.

The second method has resulted in a wide variety of digital filter algorithms which have no equivalent analogue filter. The simplest digital filters are the lowpass summing filter and the highpass difference filter, which are given respectively by:

Lowpass: $y_t = 0.5*(x_t + x_{t-1}) \rightarrow G_L(z) = 0.5*(1 + z^{-1})$

Highpass: $y_t = x_t - x_{t-1}$ $\rightarrow G_H(z) = 1 - z^{-1}$

Bandpass: $y_t = 0.5*(x_t - x_{t-2}) \rightarrow G_B(z) = G_L(z)G_H(z) = 0.5*(1 - z^{-2})$

$$... (3.54)$$

However, these simple digital filters have poor frequency response characteristics, particularly in the vicinity of the cutoff frequency, and they are presented here by way of example only.

A more important class of digital filter with sharper roll-off and better corner frequency characteristics is the *finite impulse response* (FIR) filter, the order of which is dependent on the accuracy of the results required. The form of FIR digital filters is given by

$$G(z) = z^{-M}[g_0 + \sum_{m=1}^{M} g_m(z^m + z^{-m})] \qquad ... (3.55)$$

where the weights $g_m$ are given by, noting equation (3.49),

$$g_m = \frac{1}{2\pi} \int_{-\pi}^{\pi} G(\omega)e^{-j\omega m} d\omega \qquad ... (3.56)$$

*Example: Lowpass FIR Filter:* Design a fifth order ($M = 5$) lowpass filter with cutoff frequency = 200 Hz and sampling frequency = 1 kHz ($f_N = 500$ Hz$\rightarrow$ $200\pi/500 = 0.4\pi$).

$$g_m = \frac{1}{2\pi} \int_{-\pi}^{\pi} G(\omega)e^{-j\omega m} d\omega = \frac{1}{2\pi} \int_{-0.4\pi}^{0.4\pi} 1 \times e^{-j\omega m} d\omega$$

$$= \frac{\sin(0.4\pi m)}{\pi m} = 0.4\sin c(0.4\pi m) \qquad ... (3.57)$$

where $m = 0,1,..,5$. The required lowpass FIR digital filter is then

$$G(z) = z^{-5}[g_0 + \sum_{m=1}^{5} g_m(z^m + z^{-m})]$$

$$= g_5 + g_4 z^{-1} + g_3 z^{-2} + g_2 z^{-3} + g_1 z^{-4} \qquad ... (3.58)$$

$$+ g_0 z^{-5} + g_1 z^{-6} + g_2 z^{-7} + g_3 z^{-8} + g_4 z^{-9} + g_5 z^{-10}$$

## 3.4.4 Envelope Detection

It is sometimes desirable to extract information from the *envelope* of a signal rather then the signal itself. An example of this is the case study in Chapter 9 involving the detection of boiling in nuclear reactors.

To obtain the envelope of a signal, it is first transformed into a *bandpass signal* by filtering with a narrow bandpass filter around a central frequency $f_C (= f_H = f_L)$ such that its frequency content is concentrated in a narrow band of frequencies. The resulting analogue signal $x(t)$ can be represented in general as

$$x(t) = A(t)\cos[2\pi f_C t + \phi(t)] \qquad \text{... (3.59)}$$

where $A(t)$ is the *amplitude* or *envelope* of the signal and $\phi(t)$ is the *phase* of the signal. Expanding the cosine function in (3.59) yields the alternative expression:

$$x(t) = A(t)[\cos\phi(t)\cos(2\pi f_C t) - \sin\phi(t)\sin(2\pi f_C t)]$$
$$= u_c(t)\cos(2\pi f_C t) - u_s(t)\sin(2\pi f_C t) \qquad \text{... (3.60)}$$

where

$$u_c(t) = A(t)\cos\phi(t); \quad u_s(t) = A(t)\sin\phi(t) \qquad \text{... (3.61)}$$

The in-phase quadrature (perpendicular) components $\cos(2\pi f_C t)$ and $\sin(2\pi f_C t)$ represent two rotating phasors $90^\circ$ apart, and the envelope functions, equation (3.61), are the *in-phase* and *quadrature* components of $x(t)$. As a further development of equation (3.60), the bandpass signal $x(t)$ can also be obtained from the *complex envelope*

$$u(t) = u_c(t) + ju_s(t) \qquad \text{... (3.62)}$$

so that

$$x(t) = \text{Re}[u(t)e^{j2\pi f_C t}] \qquad \text{... (3.63)}$$

In the frequency domain, we obtain by Fourier transformation

$$X(f) = \int_{-\infty}^{\infty} x(t)e^{-j2\pi ft}dt = \int_{-\infty}^{\infty} \text{Re}[u(t)e^{-j2\pi f_C t}]e^{-j2\pi ft}dt$$

$$= \frac{1}{2}\int_{-\infty}^{\infty} u(t)e^{-j2\pi(f-f_C)t}dt + \frac{1}{2}\int_{-\infty}^{\infty} u^*(t)e^{-j2\pi(f+f_C)t}dt \qquad \text{... (3.64)}$$

$$= \frac{1}{2}[U(f - f_C) + U(-f - f_C)]$$

where the identity $\text{Re}[u(t)] = [u(t) + u^*(t)]/2$ has been invoked.

Equation (3.64) shows that the spectrum of the bandpass signal $x(t)$ can be obtained from the complex signal $u(t)$ by a frequency translation, and implies that *any bandpass signal $x(t)$ can be represented by a lowpass signal $u(t)$.*

Note that, in general, $u(t)$ is complex while the bandpass signal $x(t)$ is real. In the frequency domain, $U(f)$ is an *odd* function centred at the origin ($f = 0$), whereas $X(f)$ is an *even* function centred on both sides of the origin at frequency $\pm f_c$.

It follows that the transfer function relationship $G(f)$ between two bandpassed signals $x(t)$ and $y(t)$ is, using the usual definition, $Y(f) = G(f)X(f)$. The equivalent transfer function relationship between the two (complex) lowpassed signals $u(t)$ and $v(t)$ is $V(f) = H(f)U(f)$, where $G(f)$ and $H(f)$ are related by, using equation (3.64),

$$G(f) = H(f - f_c) + H(-f - f_c) \qquad \text{... (3.65)}$$

Similarly, the equivalent lowpass impulse response $h(t)$ is given by the convolution integral

$$v(t) = \int_{-\infty}^{\infty} h(\tau)u(t - \tau)d\tau \qquad \text{... (3.66)}$$

## 3.5 FUNCTION ESTIMATES FROM FINITE LENGTH RECORDS

### 3.5.1 Bias, Consistency and Confidence Limits of Estimates

In this section we look at the estimation of functions in the time and frequency domain from finite length data sequences. We have already touched on this in section 3.3 during our discussion on discrete Fourier transforms.

In this section we are concerned with the accuracy of estimating a quantity $\theta$ associated with a random process (say $\{x_n\}$) from a limited number of samples. We would like to know how good the *estimate* $\hat{\theta}$ is compared with the true value $\theta$. The difficulty is we do not know what the true value is because it depends on averaging over *all* time. To overcome this problem, we introduce the concepts of *bias*, *consistency* and *confidence limits*.

- **Bias**: in general, the estimate $\hat{\theta}$ will be a random variable, so that the expectation of $\hat{\theta}$ is:

$$E\{\hat{\theta}\} = \theta + b \qquad \text{... (3.67)}$$

 where $b$ denotes the *bias* of the estimate. Ideally, we would like to have *unbiased estimates* ($b = 0$), and while the true value is unknowable, we can often use equation (3.67) to theoretically prove that $b$ will be zero.
- **Consistency**: this property reflects the fact that the estimate $\hat{\theta}$ should get better as the number of samples ($N$) increases, and this is expressed by

the probability

$$\lim_{N\to\infty} P\{|\hat{\theta} - \theta| \geq \Sigma\} = 0 \qquad \qquad \dots (3.68)$$

for any $\Sigma > 0$. A sufficient condition for (3.68) to be true is

$$\lim_{N\to\infty} E\{(\hat{\theta} - \theta)^2\} = 0 \qquad \qquad \dots (3.69)$$

- **Confidence Limits**: this property is based on the fact that, since the estimate $\hat{\theta}$ is a random variable, it will have an associated *probability density function* $f\{\hat{\theta}\}$, and this provides a basis for defining the probability condition:

$$\text{Prob}\{\theta - \Delta < \hat{\theta} < \theta + \Delta\} = \alpha \qquad \qquad \dots (3.70)$$

where $(\theta - \Delta)$ and $(\theta + \Delta)$ are the $100\alpha\,\%$ confidence limits for the estimate $\hat{\theta}$, and $\alpha$ is the area under the pdf distribution defined by $\theta \pm \Delta$.

### *Example*: *Estimates of the Mean Value*

Consider the estimator $\hat{\bar{x}}$ of the mean value of the random variable $\{x_n\}$:

$$\hat{\bar{x}} = \frac{1}{N}\sum_{n=0}^{N-1} x_n \qquad \qquad \dots (3.71)$$

To determine the bias we have

$$E\{\hat{\bar{x}}\} = E\left\{\frac{1}{N}\sum_{n=0}^{N-1} x_n\right\} = \frac{1}{N}\sum_{n=0}^{N-1} E\{x_n\} = \frac{1}{N}N\bar{x} = \bar{x} \qquad \dots (3.72)$$

that is, the estimator $\hat{\bar{x}}$ is *unbiased*. To determine the variance (a measure of consistency), we compute the mean square error of the mean value estimate $\hat{\bar{x}}$:

$$E\{(\hat{\bar{x}} - \bar{x})^2\} = E\left\{\left(\frac{1}{N}\sum_{n=0}^{N} x_n - \bar{x}\right)^2\right\} = \frac{1}{N^2}E\left\{\left(\sum_{n=0}^{N} x_n - N\bar{x}\right)^2\right\} = \frac{1}{N^2}E\left\{\left(\sum_{n=0}^{N}(x_n - \bar{x})\right)^2\right\}$$

and if we assume that the $\{x_n\}$ are independent such that product terms $\{x_i x_j\}$ are zero unless $i = j$ (not always true in practice) then

$$E\{(\hat{\bar{x}} - \bar{x})^2\} = \frac{1}{N^2}E\left\{\left(\sum_{n=0}^{N}(x_n - \bar{x})\right)^2\right\} = \frac{1}{N^2}(N\sigma_x^2) = \frac{\sigma_x^2}{N} \qquad \dots (3.73)$$

so that the estimator $\hat{\bar{x}}$ is consistent. The probability that the estimator $\hat{\bar{x}}$

deviates by more than $\varepsilon$ from the true mean $\bar{x}$ is given by the Chebyshev inequality

$$\text{Prob}(|\hat{\bar{x}} - \bar{x}| > \varepsilon) \leq \frac{\sigma_x^2}{N\varepsilon^2} \qquad \text{... (3.74)}$$

and hence goes to zero as $N$ goes to infinity. We will now apply these concepts to the estimates of correlation and spectral functions.

### 3.5.2   Estimates of Correlation Functions

Recall from equations (2.17), (2.26) and (2.27) that the correlation functions for infinite length records are:

$$R_{xx}(\lambda) = \int_{-\infty}^{\infty} x(t)x(t + \lambda)dt$$

$$R_{yy}(\lambda) = \int_{-\infty}^{\infty} y(t)y(t + \lambda)dt \qquad \text{... (3.75)}$$

$$R_{xy}(\lambda) = \int_{-\infty}^{\infty} x(t)y(t + \lambda)dt$$

As discussed above, *bias* or *truncation errors* arises if $x(t)$ is only defined in the range $0 \leq t \leq T$. In effect, the infinite variable $x(t)$ is multiplied by a *data window* $w(t)$ such that:

$$x_T(t) = x(t)w(t) \qquad \text{... (3.76)}$$

where

$$w(t) = 1 \quad \text{for} \quad t \leq |T|$$
$$= 0 \quad \text{for} \quad t \geq |T| \qquad \text{... (3.77)}$$

The *data window* given by equation (3.77) is termed a *rectangular window*, and in practice there is a wide variety of window functions employed to minimise the *leakage* associated with the truncation of a data record. The effects of leakage will become more apparent when we deal with the spectral functions of finite length records later.

Returning to the correlation functions, equations (3.75), substitution of (3.76) yields:

$$\bar{R}_{xx}(\tau) = \frac{1}{(T-|\tau|)} \int_0^{T-|\tau|} x_T(t)x_T(t+\tau)dt$$

$$\bar{R}_{yy}(\tau) = \frac{1}{(T-|\tau|)} \int_0^{T-|\tau|} y_T(t)y_T(t+\tau)dt \qquad \text{... (3.78)}$$

$$\bar{R}_{xy}(\tau) = \frac{1}{(T-|\tau|)} \int_0^{T-|\tau|} x_T(t)y_T(t+\tau)dt$$

For sampled data, the discrete equivalent of equations (3.78) are, with $\tau = m\Delta$ and $-M \le m \le M$:

$$\hat{R}_{xx}(m) = \frac{1}{(N-|m|)} \sum_{k=0}^{N-|m|-1} x_k \, x_{k+m}$$

$$\hat{R}_{yy}(m) = \frac{1}{(N-|m|)} \sum_{k=0}^{N-|m|-1} y_k \, y_{k+m} \qquad \text{... (3.79)}$$

$$\hat{R}_{xy}(m) = \frac{1}{(N-|m|)} \sum_{k=0}^{N-|m|-1} x_k \, y_{k+m}$$

where the summation limit arises from the fact that there are only $(N-|m|-1)$ non-zero terms in the sum on the right hand side, since $\{x_n\}$ is zero outside the range $0 \le n \le N-1$. Equations (3.79) are the *unbiased* form of the truncated estimates, since

$$E\{\hat{R}_{xx}(m)\} = E\left\{ \frac{1}{(N-|m|)} \sum_{n=0}^{N-|m|-1} x_n x_{n+m} \right\} = \frac{1}{(N-|m|)}(N-|m|) E\{x_n x_{n+m}\} = R_{xx}(m)$$

$$\text{... (3.80)}$$

Many texts give an alternative form of the correlation estimators as:

$$\hat{R}'_{xx}(m) = \frac{1}{N} \sum_{k=0}^{N} x_k \, x_{k+m}; \quad \hat{R}'_{yy}(m) = \frac{1}{N} \sum_{k=0}^{N} y_k \, y_{k+m}; \quad \hat{R}'_{xy}(m) = \frac{1}{N} \sum_{k=0}^{N} x_k \, y_{k+m}$$

$$\text{... (3.81)}$$

and it can be seen from the above discussion that these are *biased* estimators, but are an acceptable approximation (i.e. negligible bias) for data sequences where $N \gg M$.

It is not as straightforward, however, to show that the correlation estimates are also *consistent*, but they are nevertheless. Not surprisingly, the variance of correlation estimators becomes very large for $m \to N$. Also, under certain circumstances, the variance associated with the estimators using equations

(3.81) will have less variance (hence mean square error) than those obtained from equations (3.80), and will be more consistent. However, both estimators improve for $N \gg M$.

### 3.5.3 Estimates of Spectral Functions

We begin by considering the estimation of the *power spectral density* (PSD) from the sample sequence $\{x_n\}$ with $0 \le n \le N-1$. The Fourier transform of the (biased) autocorrelation estimator $\hat{R}'_{xx}(m)$, equations (3.81), yields the *periodogram* function:

$$P_{XX}(f) = \sum_{m=-(N-1)}^{N-1} \hat{R}'_{xx}(m)e^{-j2\pi fm} = \sum_{m=-(N-1)}^{N-1}\left[\frac{1}{N}\sum_{n=0}^{N-1}x_n x_{n+m}e^{-j2\pi fm}\right]$$

$$= \frac{1}{N}\sum_{n=0}^{N-1}x_n e^{j2\pi fn}\sum_{k=0}^{N-1}x_k e^{-j2\pi fk} = \frac{1}{N}X^*(f)X(f) = \frac{|X(f)|^2}{N} \qquad \dots (3.82)$$

where $k = n + m$ and we have again invoked the fact that $\{x_n\}$ is zero outside the range $0 \le n \le N-1$ to change the summation limits. The final result follows from our definition of $X(f)$ given in equation (3.36).

The periodogram was originally conceived for the purpose of highlighting suspected periodicities in meteorological data. The periodogram may be computed by the DFT and FFT methods discussed earlier. We now investigate the *bias* of the estimator $P_{xx}(f)$ by taking the expectation of (3.82):

$$E\{P_{XX}(f)\} = E\left\{\frac{1}{N}\sum_{m=-(N-1)}^{N-1}\hat{R}'_{xx}(m)e^{-j2\pi fm}\right\} = \frac{1}{N}\sum_{m=-(N-1)}^{N-1}E\{\hat{R}'_{xx}(m)\}e^{-j2\pi fm}$$

that is

$$E\{P_{XX}(f)\} = \sum_{m=-(N-1)}^{N-1}\left[\frac{1}{N^2}\sum_{n=0}^{N-1}E\{x_n x_{n+m}\}e^{-j2\pi fm}\right] = \sum_{m=-(N-1)}^{N-1}\left[\frac{1}{N^2}N(N-|m|)R_{xx}(m)e^{-j2\pi fm}\right]$$

$$= \sum_{m=-(N-1)}^{N-1}\left[\left(1-\frac{|m|}{N}\right)R_{xx}(m)e^{-j2\pi fm}\right]$$

$$\dots (3.83)$$

where the penultimate relationship follows from equation (3.80). Equation (3.83) shows that the mean of the estimated spectrum is the Fourier transform of the 'true' autocorrelation function $R_{xx}(m)$ multiplied by a triangular *lag window* function $w(m)$, which is equivalent to the convolution of the rectangular data window $w(t)$, equation (3.77), with itself; that is,

$$w(m) = E\{w_n w_{n+m}\} = \frac{1}{(N-|m|)} \sum_{n=0}^{N-|m|-1} w_n w_{n+m}$$

$$= \left(1 - \frac{|m|}{M}\right) \qquad |m| \le M \qquad \qquad \text{... (3.84)}$$

$$= 0 \qquad \qquad |m| > M$$

where $M$ is the maximum number of lags (unrelated to the binary exponent in equation (3.40)). Equation (3.84) is known as the *Bartlett lag window* and the corresponding *Bartlett spectral window* is obtained by Fourier transformation of (3.84) to yield:

$$W_B(f) = M\left(\frac{\sin(\pi f M)}{\pi f M}\right)^2 = M\mathrm{sinc}^2(\pi f M) \qquad \text{... (3.85)}$$

As noted previously, a wide variety of data, lag and spectral windows have been developed to minimise the *leakage* in the spectral estimates, and we will consider these in more detail in the next chapter. In the present context we note that the periodogram is a convolution of the 'true' (theoretical) spectral density $S_{XX}(f)$ and the Bartlett spectral window; that is, in continuous time

$$E[P_{XX}(f)] \approx \int_{-\infty}^{\infty} S_{XX}(\alpha)W_B(f-\alpha)d\alpha = \overline{S}_{XX}(f) \qquad \text{... (3.86)}$$

where $\overline{S}_{XX}(f)$ is the 'smoothed' (windowed) power spectral density. Hence the bias in the power spectral density estimator can be defined by

$$B(f) = E[P_{XX}(f)] - S_{XX}(f) = \overline{S}_{XX}(f) - S_{XX}(f)$$

$$= E\left[\int_{-\infty}^{\infty} w(\tau)R_{xx}(\tau)e^{-j2\pi f\tau}d\tau\right] - \int_{-\infty}^{\infty} R_{xx}(\tau)e^{-j2\pi f\tau}d\tau \qquad \text{... (3.87)}$$

$$\approx \int_{-\infty}^{\infty} (w(\tau)-1)R_{xx}(\tau)e^{-j2\pi f\tau}d\tau$$

From equation (3.84), it follows that for a Bartlett window the bias is

$$B_B(f) \approx \frac{1}{M}\int_{-\infty}^{\infty} -|\tau|R_{xx}(\tau)e^{-j2\pi f\tau}d\tau \qquad \text{... (3.88)}$$

and is *small* when the number of lags $M$ is *large*. This means that the bias can only be made small by making the spectral window $W(f)$ narrow or as close to a delta function as possible.

However, for consistent estimates, the *variance* of the spectral estimators, equation (3.69), must also be small. If it is assumed the process is Gaussian, then it can be shown that

$$\lim_{N \to \infty} \text{Var}[P_{XX}(f)] \approx S_{XX}^2(f) \qquad \qquad \text{... (3.89)}$$

which says that the periodogram is not a consistent estimate of the true power spectral density spectrum $S_{XX}(f)$; that is, it does not converge to the true power density spectrum. It also follows that a *narrow* spectral window $W(f)$ will result in a *large variance*, and that some compromise procedure is needed to achieve a balance between these conflicting requirements. A sensible procedure is to make the *mean square error* of the function

$$\text{Var}[P_{XX}(f)] + B^2(f) \qquad \qquad \text{... (3.90)}$$

as small as possible for each frequency. The exact nature of the compromise will depend on the degree of smoothness of the spectral density estimator $S_{XX}(f)$. For example, if $S_{XX}(f)$ is very smooth, then a wide spectral window may be used without having a serious affect on the bias.

A practical compromise is to compute the average of a number of periodograms using the following procedure:

1) Determine the number of samples $N$ for which the signals $N$ samples apart are approximately uncorrelated (independent), that is, $E[x_n x_{n+N}]$ is small.
2) Divide the sample record into $Q$ contiguous segments (batches) of $N$ samples.
3) Compute the average periodogram

$$\overline{P}_{XX}(f) = \frac{1}{Q} \sum_{q=0}^{Q-1} P_{XX}(f,q) \qquad \qquad \text{... (3.91)}$$

where $P_{XX}(f,q)$ is the periodogram computed using (3.82) for the $q$th data segment.

With certain assumptions regarding stationarity and independence, it can be shown that the estimate $\overline{P}_{XX}(f)$ given by equation (3.91) is asymptotically unbiased and consistent as $QN \to \infty$. There are many refinements and alternative procedures that could be employed, but the above procedure is a good start.

Before leaving this section, it should be noted that the above concepts and limitations also apply to *cross-spectral density* estimation. However, the resulting bias and consistency relationships are more arduous to prove mathematically, and will not be pursued here. Interested readers are referred to the excellent text by Jenkins and Watts (1968) for further details.

It should be apparent from the above discussion that estimation theory can be quite tedious, and computing the estimates in practice often requires simplifying assumptions, the most common being *stationarity* and *normality* (i.e. Gaussian random processes). The procedures for deriving estimates based on assumptions can be disconcertingly circular. As is often the case in practice, experience is the key. A good idea of what to expect in a given application can help to get the analysis procedures started. Refinements can then be made as successive results are produced. Decisions as to whether the observed results are true or are due to leakage (say) require experience and much patience.

## REFERENCES

Jenkins, G. M. & Watts, D. G., 1968, *Spectral Analysis and its Applications*, Holden-Day (San Francisco).

Kamen, E. W., 1990, *Introduction to Signals and Systems*, Macmillan (New York).

Oppenheim, A. V., & Schafer, R. W., 1989, *Discrete Time Signal Processing*, Prentice-Hall (London).

Proakis, J. G., & Manolakis, D. G., 1992, *Digital Signal Processing: Principles, Algorithms and Applications*, Macmillan (New York).

Proakis J. G., Rader, C. M., Ling, F., & Nikias, C. L., 1992, *Advanced Digital Signal Processing*, Macmillan (New York).

Ziemer, R. E., Tranter, W. H., & Fannin, D. R., 1990, *Signals and Systems: Continuous and Discrete*, Macmillan Publishing Company (New York).

## TUTORIAL EXERCISES

1) An industrial process is modelled by the second order continuous differential equation

$$a_2 \frac{d^2 y(t)}{dt^2} + a_1 \frac{dy(t)}{dt} + a_0 y(t) = b_0 x(t)$$

a) Evaluate the coefficients when this equation is written in the form:

$$(1 + T_1 D)(1 + T_2 D) y(t) = K x(t)$$

where $T_1$ and $T_2$ are the process time constants, $K$ is the static gain and $D \equiv d/dt$ is the differential operator.

*b)* Obtain the continuous process transfer function $G(s)$ (see Appendix B3) and prove that the output response $y(t)$ to an *unit step input* at time zero $(t = 0)$ is given by:

$$y(t) = \frac{K}{T_1 - T_2}\left[1 - T_1^2 e^{-t/T_1} + T_2^2 e^{-t/T_2}\right]$$

*c)* Obtain the discrete process transfer function $G(z)$ (see Appendix B3) and prove that the output response, $y_t$, to an *unit step input* at time zero $(t = 0)$ is given by:

$$y_t = -(\alpha_1 - \alpha_0)y_{t-1} - (\alpha_2 - \alpha_1)y_{t-2} + \alpha_2 y_{t-3} + \beta_1 + \beta_2$$

and evaluate the $\alpha$'s and $\beta$'s in terms of the process time contants and static gain. Plot the continuous and discrete responses for $K=5$, $T_1 = 10$ s and $T_2 = 20$ s.

2) Compute and plot the magnitude and phase of the 16-point DFT $(N = 16, \Delta = 1)$ of the sinusoids:

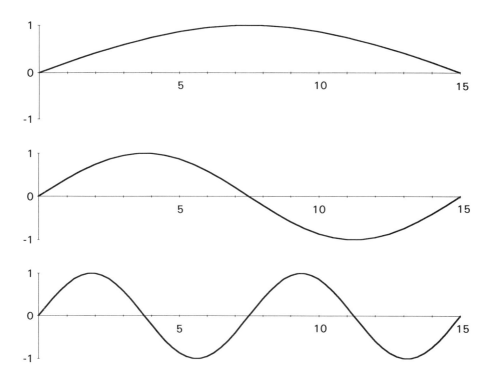

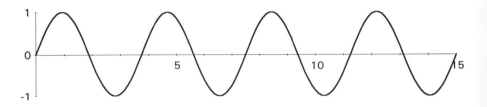

3) Use the definition of the DFT pair, equations (3.36) and (3.37) to prove the following properties:

   a) Linearity:

   $$a x_1(n) + b x_2(n) \Leftrightarrow a X_1(k) + b X_2(k)$$

   b) Convolution/Product:

   i)     $$x_1(n) \cdot x_2(n) \Leftrightarrow \frac{1}{N} \sum_{m=0}^{N-1} X_1(m) X_2(k-m)$$

   ii)    $$\sum_{m=0}^{N-1} x_1(m) x_2(n-m) \Leftrightarrow X_1(k) \cdot X_2(k)$$

4) In attempting to measure the average velocity of a fluid, two sensors are placed a known distance apart. The first sensor produces an output $x(t)$ and the second sensor produces an output given by:

   $$y(t) = x(t-\tau) + n(t),$$

   where $\tau$ is the transport delay. The inclusion of the term $n(t)$ is to account for the uncorrelated noise that may be added as the fluid is transported between the two sensors.

   a) Re-write the above equations in discrete form and use them in estimating the cross-correlation function between the two sensors for a range of signal-to-noise ratios. (Hint: you may define the signal-to-noise ratio in terms of the ratio of the appropriate autocorrelation functions evaluated at zero lag – see equation (2.18) in Chapter 2).

   b) What signal-to-noise ratio would you feel confident in using in practice?

5) Design a bandpass filter of tenth order (M=10) for a signal sampled at 10 kHz for use in the frequency band 2 kHz to 4 kHz.

6) Data is smoothed according to the algorithm

   $$y_n = 1/3\{x_n + x_{n-1} + x_{n-2}\}$$

In principle it is possible to restore the original data sequence by processing the smoothed data with an algorithm whose transfer function is the reciprocal of the original one. Show that there will be stability problems with such an algorithm.

7)  Sketch the frequency response of

$$H(z) = \frac{0.24z^{-1}}{1 - 1.16z^{-1} + 0.41z^{-2}}$$

8)  Compute the FFT of a function which is a triangular function at the origin and compare your result with the Fourier transform of the continuous signal.

9)  A signal pulse is in the form of a sinc function at the origin. Compute and plot the magnitude spectrum at sample rates below and above the Nyquist rate.

10) a)  Derive the Z transform of samples from $te^{-\alpha t}$ and

    b)  the inverse transform of $\dfrac{z^2}{(z-1)(z-0.5)}$

# APPENDIX A3

**Table A3.1** Properties of the Z-transform

| PROPERTY | EQUATION |
|---|---|
| Linearity ($\alpha$, $\beta$ scalars) | $Z\{\alpha f_k + \beta g_k\} = \alpha F(z) + \beta G(z)$ |
| Time Shifting/Real Translation | $Z\{f_{k-n}\} = z^{-n} F(z)$ |
| Time Shifting | $Z\{f_{k+1}\} = z[F(z) - f_0]$ |
| Summation | $Z\left\{\sum_{j=0}^{k} f_j\right\} = \dfrac{z}{z-1} Z\{f_k\} = \dfrac{z}{z-1} F(z)$ |
| Frequency Differentiation | $Z\{k f_k\} = -z\dfrac{d}{dz}Z\{f_k\} = -z\dfrac{dF(z)}{dz}$ |
| Frequency Integration | $Z\left\{\dfrac{1}{k} f_k\right\} = \int_z^{\infty} \dfrac{F(w)}{w} dw \; ; \; F(w) = Z\{f_k\}\big|_{z=w}$ |
| Frequency Shifting | $Z\{\sigma^k f_k\} = F(\sigma^{-1} z)$ |
| Convolution | $H(z) = G(z) F(z) = Z\left\{\sum_{j=0}^{k} f_j \, g_{k-j}\right\}$ |
| Inverse Transformation | $f_k = \dfrac{1}{2\pi j} \oint \dfrac{F(z) z^k \, dz}{z}$ <br> Note anticlockwise encirclement of the origin. |
| Damping | $Z\{f_k e^{ak\Delta}\} = F(ze^{a\Delta})$ |
| Initial Value Theorem | $f_0 = \lim_{z\to\infty}[F(z)]$ |
| Final Value Theorem | $f_\infty = \lim_{z\to 1}[(1 - z^{-1}) F(z)]$ |

# APPENDIX B3

## CONTINUOUS AND DISCRETE MODEL RELATIONSHIPS

### B3.1    Relation Between Continuous and Discrete Models

It is often desirable in some practical applications to know the parameter relationships between a continuous model for a process described by an ordinary differential equation (ODE) of the form given by equation (3.11), and its corresponding discrete process ARMA model given by equation (3.12). In the analysis outlined below, the discrete model parameters are obtained from the continuous model parameters by the following steps:

1)  Laplace transform the ODE to obtain its transfer function model $G(s)$:

$$G(s) = \frac{\sum_{k=0}^{m} b_k s^k}{\sum_{j=0}^{n} a_j s^j} \qquad \text{... (B3.1)}$$

2)  Multiply $G(s)$ by the transfer function $G_S(s)$ of the **sample and hold** device (if used) to obtain the **hold transfer function** $G_H(s)$

$$G_H(s) = G_S(s)G(s) \qquad \text{... (B3.2)}$$

3)  Use **partial fractions** to express $G_H(s)$ as a series of first order fractions (this step makes it easier to obtain the desired time response):

$$G_H(s) = \sum_{k=0}^{n} \frac{A_k}{s + p_k} \qquad \text{... (B3.3)}$$

where the coefficients $A_k$ are obtained using the identity

$$A_k = \left[(s + p_k)G_H(s)\right]_{s=-p_k} \qquad \text{... (B3.4)}$$

4)  Obtain the time response and corresponding data sequence by **inverse Laplace transformation** of $G_H(s)$.

5)  Take the Z-transform of the corresponding data sequence to obtain the discrete transfer function

$$G(z) = Z\{G_H(s)\} \qquad \text{... (B3.5)}$$

6) Evaluate the $\alpha$ and $\beta$ parameters of the difference equation by equating the coefficients of equation (3.20) with equation (B3.5) for the same powers of the $z$-operator.

**Note** : Steps (2) to (5) can be eliminated by the use of appropriate Tables giving $s \rightarrow t$ and $t \rightarrow z$ or, more directly, $s \rightarrow z$. However, it is important that the reader understand the transformation concepts and is familiar with the above procedure **before** taking short-cuts using Tables.

### B3.2    Example: Second Order Discrete System

*Step (1): Transfer Function*

Laplace transformation of the second order continuous process, equation (3.15), yields the transfer function (assuming zero initial conditions):

$$G(s) = \frac{K}{(1 + T_1 s)(1 + T_2 s)} \qquad \qquad \text{... (B3.6)}$$

*Step (2): Hold Transfer Function*

Assume **zero-order** sample and hold with unit gain over the sample interval $\Delta$, that is,

$$f(t) = 1 \qquad 0 \le t \le \Delta$$
$$= 0 \qquad t > \Delta \qquad \qquad \text{... (B3.7)}$$

Hence

$$G_S(s) = \int_0^\Delta e^{-st} dt = \left[ \frac{-e^{-st}}{s} \right]_0^\Delta = \frac{1 - e^{-s\Delta}}{s} \qquad \qquad \text{... (B3.8)}$$

The hold transfer function is given by

$$G_H(s) = G_S(s)G(s) = \frac{K(1 - e^{-s\Delta})}{s(1 + T_1 s)(1 + T_2 s)}$$

$$= \frac{K}{s(1 + T_1 s)(1 + T_2 s)} - \frac{Ke^{-s\Delta}}{s(1 + T_1 s)(1 + T_2 s)} \qquad \qquad \text{... (B3.9)}$$

*Step (3): Partial Fraction Expansion*

Expanding the first term of the above equation into partial fractions we have:

$$G'_H(s) = \frac{K}{s(1+T_1s)(1+T_2s)} = \frac{A_0}{s} + \frac{A_1}{1+T_1s} + \frac{A_2}{1+T_2s} \qquad \dots \text{(B3.10)}$$

from which, using the identity (B3.4) or by writing the right hand side of (B3.10) in the same form as the LHS and equating coefficients, we obtain:

$$A_0 = \left[sG'_H(s)\right]_{s=0} = K$$

$$A_1 = \left[(1+T_1s)G'_H(s)\right]_{s=-1/T_1} = \frac{-KT_1^2}{T_1 - T_2} \qquad \dots \text{(B3.11)}$$

$$A_2 = \left[(1+T_2s)G'_H(s)\right]_{s=-1/T_2} = \frac{KT_2^2}{T_1 - T_2}$$

*Step (4): Time Response*

Taking the inverse Laplace transform of the individual terms of equation (B3.11) yields:

$$\mathcal{L}^{-1}\left\{\frac{A_0}{s}\right\} = K$$

$$\mathcal{L}^{-1}\left\{\frac{A_1}{1+T_1s}\right\} = \frac{-KT_1e^{-t/T_1}}{T_1 - T_2} \qquad \dots \text{(B3.12)}$$

$$\mathcal{L}^{-1}\left\{\frac{A_1}{1+T_2s}\right\} = \frac{KT_2e^{-t/T_2}}{T_1 - T_2}$$

Hence

$$\mathcal{L}^{-1}\{G'_H(s)\} = K\left[1 - \frac{T_1e^{-t/T_1}}{T_1 - T_2} + \frac{T_2e^{-t/T_2}}{T_1 - T_2}\right] \qquad \dots \text{(B3.13)}$$

This corresponds to the data sequence

$$g(k\Delta) = K\left[1 - \frac{T_1e^{-k\Delta/T_1}}{T_1 - T_2} + \frac{T_2e^{-k\Delta/T_2}}{T_1 - T_2}\right] \qquad \dots \text{(B3.14)}$$

*Step (5): Discrete Transfer Function*

Taking Z-transforms of the individual terms in equation (B3.14) we obtain

$$G'(z) = Z\{G'_H(s)\} = \frac{K}{T_1 - T_2}\left[\frac{(T_1 - T_2)z}{z - 1} - \frac{T_1z}{z - e^{-\Delta/T_1}} + \frac{T_2z}{z - e^{-\Delta/T_2}}\right] \qquad \dots \text{(B3.15)}$$

Similarly,

$$Z\{G_H''(s)\} = Z\{-e^{-s\Delta}G_H'(s)\} = z^{-1}Z\{G_H'(s)\} \qquad \text{... (B3.16)}$$

since, from equation (3.29), $z^{-1} = e^{-s\Delta}$. Hence

$$G(z) = Z\{G_H(s)\} = \frac{K}{T_1 - T_2}\left[(T_1 - T_2) - \frac{T_1(1-z^{-1})}{1-S_1 z^{-1}} + \frac{T_2(1-z^{-1})}{1-S_2 z^{-1}}\right] \qquad \text{... (B3.17)}$$

where

$$S_1 = e^{-\Delta/T_1}; \quad S_2 = e^{-\Delta/T_2} \qquad \text{... (B3.18)}$$

Converting equation (B3.17) into the form given by equation (3.20):

$$(1+\alpha_1 z^{-1} + \alpha_2 z^{-2})y_t = (\beta_0 + \beta_1 z^{-1} + \beta_2 z^{-2})x_t$$

and equating the numerator and denomenator coefficients yields, respectively, the moving average ($\beta$) and autoregressive ($\alpha$) coefficients:

$$z^0 \rightarrow \beta_0 = 0$$

$$z^{-1} \rightarrow \beta_1 = \frac{K}{T_1 - T_2}\left[T_1(1-S_1) - T_2(1-S_2)\right] \qquad \text{... (B3.19)}$$

$$z^{-2} \rightarrow \beta_2 = \frac{K}{T_1 - T_2}\left[T_1(1-S_1)S_2 - T_2(1-S_2)S_1\right]$$

and

$$\alpha_1 = -(S_1 + S_2); \quad \alpha_2 = S_1 S_2 \qquad \text{... (B3.20)}$$

Hence the discrete time response (Figure B3.1) is given by

$$y_t = -\alpha_1 y_{t-1} - \alpha_2 y_{t-2} + \beta_1 x_{t-1} + \beta_2 x_{t-2} \qquad \text{... (B3.21)}$$

The response of this process to (say) a step input will depend on whether the roots are complex ($C$) or real ($R$), as shown in Figure B3.1.

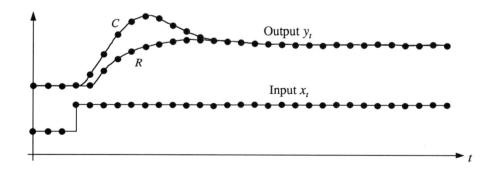

**Figure B3.1** Second order discrete system response to a step input

# 4

# BIVARIATE LINEAR PROCESS ANALYSIS

## 4.1    INTRODUCTION AND LEARNING OBJECTIVES

This chapter presents some of the more important aspects of spectral analysis theory and its application to the analysis of *bivariate linear processes*. Bivariate linear processes are the basic building blocks for analysing the more complex higher order multivariate processes discussed in Chapter 7.

In this Chapter, we show how to:

- calculate the minimum least squares transfer function estimates for bivariate linear processes;
- define and compute the *ordinary* coherence function for bivariate linear processes;
- define and compute the *confidence limits* of the computed transfer function estimates;
- analyse bivariate processes with extraneous (uncorrelated) noise on the input and output;
- analyse the effects of *signal preprocessing* operations (data filtering, data smoothing for leakage reduction, data alignment compensation for process time delays) on the computed transfer function estimates.

## 4.2    BIVARIATE PROCESS IDENTIFICATION

### 4.2.1    SISO Transfer Function Models

A bivariate linear process is one in which the output is linearly related to a single input (Figure 4.1), and is often termed a single-input-single-output (SISO) process. The models of SISO processes give a measure of the *transfer relationship* between the input and the output in either the time domain (weighting functions) or frequency domain (transfer functions). Only the identification of transfer function models is discussed in this chapter, but it should be noted that, as we have seen in Chapter 2, the weighting functions and the transfer functions are, theoretically at least, *Fourier transform pairs*.

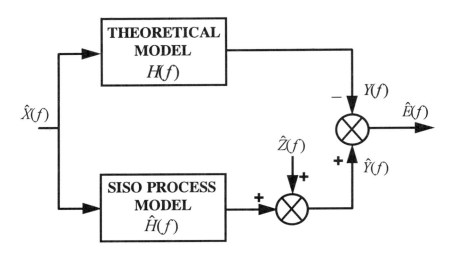

**Figure 4.1** Block diagram of SISO process and theoretical models

Referring to Figure 4.1, we wish to know *firstly* how accurate the SISO theoretical model $H(f)$ is in modelling the process and *secondly* how reliable the statistically based SISO process model estimates $\hat{H}(f)$ are in measuring process characteristics. For comparison purposes, we assume both models are 'driven' by the same measurement of the process input $\hat{X}(f)$.

In practice, the process measurements are usually contaminated by *extraneous noise sources* which degrade the reliability of the spectral estimates from which the transfer function estimates $\hat{H}(f)$ are derived, and these are 'lumped' together as an unknown variable $\hat{Z}(f)$. In order to achieve some degree of compatibility between the *theoretical model* $H(f)$ and the *process model* $\hat{H}(f)$ over the specified frequency bandwidth, it is necessary to minimise the residual error function $\hat{E}(f)$ and derive the relationships between the functions $H(f)$ and $\hat{H}(f)$ and between $\hat{E}(f)$ and $\hat{Z}(f)$. (Note that the hat (^) above the symbol denotes an *estimate* of the true value.)

The residual mean square error over all frequencies is defined by:

$$\overline{\hat{e}^2} = \mathrm{E}[\hat{e}^2(t)] = \hat{R}_{ee}(0) = \int_{-\infty}^{\infty} \hat{S}_{EE}(f)df \qquad \text{... (4.1)}$$

where $\hat{S}_{EE}(f)$ is the power spectral density estimate of the residual error

$\hat{E}(f)$, and is given by

$$
\begin{aligned}
\hat{S}_{EE}(f) &= \lim_{T \to \infty} \mathrm{E}[\hat{E}^*(f)\hat{E}(f)] \\
&= \lim_{T \to \infty} \mathrm{E}[\{\hat{Y}^*(f) - Y(f)\}\{\hat{Y}(f) - Y(f)\}] \\
&= \lim_{T \to \infty} \mathrm{E}[\{\hat{Z}^*(f) + \hat{X}^*(f)\hat{H}^*(f) - \hat{X}^*(f)H^*(f)\} \\
&\qquad\qquad \{\hat{Z}(f) + \hat{X}(f)\hat{H}(f) - \hat{X}(f)H(f)\}] \\
&= \hat{S}_{ZZ}(f) + \\
&\quad \hat{S}_{XX}(f)\left[\left|\hat{H}(f)\right|^2 - \hat{H}^*(f)H(f) - H^*(f)\hat{H}(f) + |H(f)|^2\right]
\end{aligned}
$$

$$... (4.2)$$

in which we have assumed that the input $\hat{X}(f)$ and the extraneous noise $\hat{Z}(f)$ are uncorrelated; that is, the cross-product terms $\mathrm{E}[\hat{X}^*(f)\hat{Z}(f)]$ and $\mathrm{E}[\hat{Z}^*(f)\hat{X}(f)]$ are zero.

The residual mean square error, equation (4.1), will be a minimum when:

$$
\frac{\partial \hat{S}_{ZZ}(f)}{\partial \hat{H}^*(f)} = \hat{S}_{XX}(f)\left[\hat{H}(f) - H(f)\right] = 0 \qquad ... (4.3)
$$

and since $\hat{S}_{XX}(f) \neq 0$ in general, it follows that $\overline{e^2}$ will be a minimum when the theoretical transfer function $H(f)$ is identically equivalent to the transfer function estimate $\hat{H}(f)$ at all frequencies; that is,

$$
H(f) \equiv \hat{H}(f). \qquad ... (4.4)
$$

On substitution of this result into equation (4.2), we also note that $\overline{e^2}$ will be a minimum when

$$
\hat{S}_{EE}(f) = \hat{S}_{ZZ}(f). \qquad ... (4.5)
$$

Equations (4.4) and (4.5) are what we would expect intuitively; namely, the accuracy of theoretical model $H(f)$ is heavily dependent on the statistical reliability of the process model estimates $\hat{H}(f)$, and these in turn are dependent on the magnitude and frequency characteristics of the extraneous noise term $\hat{Z}(f)$. Their significance will become more apparent in the following sections, where particular attention is given to:

1) the minimum least squares estimation of the process transfer functions $\hat{H}(f)$;

2) the statistical confidence limits of the transfer function estimates $\hat{H}(f)$;

3) the analysis of processes with extraneous (uncorrelated) noise on the input and output;
4) the data filtering, smoothing and alignment procedures for minimising bias in the computed spectral (hence transfer function) estimates.

### 4.2.2    Stationarity and Ergodicity Assumptions

The analysis of SISO and higher order processes assume that they can be identified from *stationary* and *ergodic* realizations of finite length $T$ $(T \to \infty)$. The input and output time series are said to be *weakly stationary* when the mean, variance and correlation functions are independent of the time $(t_i)$ of commencement of the sample record, that is,

$$\hat{\bar{x}}(t_i) = \hat{\bar{x}}; \quad \hat{\sigma}^2(t_i) = \hat{\sigma}^2; \quad \hat{R}(t_i, t+\tau) = \hat{R}(\tau)$$

Stationary time series are said to be *weakly ergodic* if the mean, variance and correlation functions are equal to their corresponding ensemble averaged functions. Thus the time series is weakly ergodic if, for the $k$th sample function,

$$\hat{\bar{x}}(k) = \hat{\bar{x}}; \quad \hat{\sigma}^2(k) = \hat{\sigma}^2; \quad \hat{R}(\tau, k) = \hat{R}(\tau)$$

A summary of basic relationships for stationary processes and their respective symbols is given in Appendix A4.

### 4.3        MINIMUM LEAST SQUARES TRANSFER FUNCTION ANALYSIS

### 4.3.1    Physical Realizability Condition

Consider a process with input and output realizations $\hat{x}(t)$ and $\hat{y}(t)$ respectively, and an *extraneous* or *residual* noise component $\hat{z}(t)$ superimposed on the output (Figure 4.1). The noise term $\hat{z}(t)$ is assumed to be an *error term* which contains a *systematic component* due to the inadequacy of the linear process approximation, and a *random component* due to measurement errors and other uncontrolled variables affecting the output. The output is calculated from a weighted average of the input with the weighting function $\hat{h}(\tau)$ according to the convolution integral

$$\hat{y}(t) - \hat{\bar{y}} = \int_0^\infty \{\hat{x}(t-\tau) - \hat{\bar{x}}\}\,\hat{h}(\tau)\,d\tau + \hat{z}(t) \qquad \text{... (4.6)}$$

where $\tau$ is a *time shift* operator. The lower (zero) limit of integration assumes that, for any physical process, the weighting function $\hat{h}(\tau)$ must be zero for negative (lead) values of $\tau$, which means that the process cannot respond to inputs it has not yet received; that is, the process must satisfy the *physical realizability condition*,

$$\hat{h}(\tau) = 0; \quad \tau < 0 \qquad \qquad \text{... (4.7)}$$

The mean values of $\hat{x}(t)$ and $\hat{y}(t)$ in equation (4.6) are defined as (see Appendix A4)

$$\hat{\bar{x}} = E[\hat{x}(t)] = \int_0^T \hat{x}(t)\,dt$$
$$\qquad \qquad \text{... (4.8)}$$
$$\hat{\bar{y}} = E[\hat{y}(t)] = \int_0^T \hat{y}(t)\,dt$$

and may be assumed, without loss of generality, to be zero.

### 4.3.2    Minimum Least Squares Transfer Function Estimation

Fourier transformation of equation (4.6) yields the equivalent frequency domain equation:

$$\hat{Y}(f) = \hat{X}(f)\hat{H}(f) + \hat{Z}(f) \qquad \qquad \text{... (4.9)}$$

where the *process transfer function* $\hat{H}(f)$ forms a Fourier transform pair with the process weighting function $\hat{h}(\tau)$. Ideally, $\hat{H}(f)$ should be the '*optimum*' transfer function in the minimum least squares sense; that is, the necessary condition under which the residual noise term, $\hat{z}(t)$ or $\hat{Z}(f)$, is a minimum. This is proved in the following analysis.

As shown in section 4.2.1, the mean square error in the estimation of the transfer function, $\hat{H}(f)$, over all frequencies is given by

$$\overline{\hat{z}^2} = E[\hat{z}^2(t)] = \hat{R}_{zz}(0) = \int_{-\infty}^{\infty} \hat{S}_{zz}(f)\,df \qquad \qquad \text{... (4.10)}$$

where $\hat{S}_{zz}(f)$ is the power spectral density function of the residual noise $\hat{Z}(f)$, and is defined by

$$\hat{S}_{zz}(f) = \lim_{T \to \infty} E[\hat{Z}^*(f)\hat{Z}(f)]$$
$$\qquad \qquad \text{... (4.11)}$$
$$= \lim_{T \to \infty} E[\{\hat{Y}^*(f) - \hat{X}^*(f)\hat{H}^*(f)\}\{\hat{Y}(f) - \hat{X}(f)\hat{H}(f)\}]$$

that is,

$$\hat{S}_{ZZ}(f) = \hat{S}_{YY}(f) - \hat{S}_{YX}(f)\hat{H}(f) - \hat{S}_{XY}(f)\hat{H}^*(f) + \hat{S}_{XX}(f)\left|\hat{H}(f)\right|^2$$

$$... (4.12)$$

Substitution of equation (4.12) into (4.10) yields

$$\overline{\hat{z}^2} = \overline{\hat{y}^2} - \int_{-\infty}^{\infty} [\hat{S}_{YX}(f)\hat{H}(f) - \hat{S}_{XY}(f)\hat{H}^*(f) + \hat{S}_{XX}(f)\left|\hat{H}(f)\right|^2] df$$

$$... (4.13)$$

In order to minimise equation (4.8), $\hat{H}(f)$ and its complex conjugate $\hat{H}^*(f)$ are expressed in terms of their orthogonal real and imaginary components as follows:

$$\hat{H}(f) = \hat{A}(f) - j\hat{B}(f)$$
$$\hat{H}^*(f) = \hat{A}(f) + j\hat{B}(f)$$

$$... (4.14)$$

Substitution of equation (4.14) into (4.13) yields:

$$\overline{\hat{z}^2} = \overline{\hat{y}^2} - \int_{-\infty}^{\infty} [\hat{S}_{YX}(\hat{A} - j\hat{B}) - \hat{S}_{XY}(\hat{A} + j\hat{B}) + \hat{S}_{XX}(\hat{A}^2 + \hat{B}^2)] df$$

where the frequency dependence of the functions under the integral have been omitted for simplicity. Differentiation of this equation with respect to $\hat{A}(f)$ and $\hat{B}(f)$ yields:

$$\frac{\partial}{\partial \hat{A}}(\overline{\hat{z}^2}) = \hat{S}_{YX} + \hat{S}_{XY} - 2\hat{S}_{XX}\hat{A} = 0$$

$$... (4.15)$$

$$\frac{\partial}{\partial \hat{B}}(\overline{\hat{z}^2}) = -\hat{S}_{YX} + \hat{S}_{XY} + j2\hat{S}_{XX}\hat{B} = 0$$

Adding equations (4.15) gives the minimum mean square transfer function estimate $\hat{H}(f)$ in terms of the measured cross spectral and power spectral density estimates as

$$\hat{S}_{XY}(f) = \hat{S}_{XX}(f)\hat{H}(f)$$

$$... (4.16)$$

(Note that this result may be obtained directly from equation (4.13) by differentiating it with respect to $\hat{H}^*(f)$ as in section 4.2.1, but we have taken this longhand approach here as a precursor to a proof given in Chapter 7 for multivariate processes.)

Inverse Fourier Transformation of equation (4.16) yields the equivalent Wiener-Hopf integral equation

$$\hat{R}_{xy}(\tau) = \int_{-\infty}^{\infty} \hat{R}_{xx}(\tau - \lambda)\hat{h}(\lambda) d\lambda$$

$$... (4.17)$$

This important relationship may be proved alternatively using the calculus of variations (Jenkins and Watts 1968).

It could be noted that premultiplication of equation (4.9) by the complex conjugate $\hat{X}^*(f)$ and taking expectations gives

$$\hat{S}_{XY}(f) = \hat{S}_{XX}(f)\,\hat{H}(f) + \hat{S}_{XZ}(f) \qquad \ldots (4.18)$$

and comparing equations (4.16) and (4.18) it follows that:

$$\hat{S}_{XZ}(f) = 0 \qquad \ldots (4.19)$$

for *all* frequencies in order for equation (4.16) to be satisfied. As we have seen, the condition (4.19) requires the input signal $\hat{x}(t)$ and the residual noise $\hat{z}(t)$ to be uncorrelated (independent); that is, that $\hat{z}(t)$ is *white* noise. Priestly (1981) investigated the more general case where $\hat{z}(t)$ is *coloured* noise, and derives alternative conditions for equations (4.16) and (4.19) (see Chapter 5).

## 4.4 ORDINARY COHERENCE FUNCTION

An important measure of the *coherent* information transferred between the process input and output is given by the *ordinary* coherence function, which is defined as:

$$\hat{\gamma}_{XY}^2(f) = \frac{\left|\hat{S}_{XY}(f)\right|^2}{\hat{S}_{XX}(f)\hat{S}_{YY}(f)} \qquad \ldots (4.20)$$

(The term *ordinary* coherence function is used to denote that this relationship applies to bivariate processes only, and to differentiate it from the *partial* and *multiple* coherence functions defined for higher order multivariate processes in Chapter 7.)

For an *ideal* (perfectly coherent) process $\hat{\gamma}^2(f) = 1$, although in practice it is possible to obtain values greater than unity if either $\hat{S}_{XX}(f)$ or $\hat{S}_{YY}(f)$ is very small at certain frequencies. When the input and output time series are uncorrelated (incoherent) $\hat{\gamma}^2(f) = 0$. In general the ordinary coherence function lies in the range $0 \le \hat{\gamma}_{XY}^2(f) \le 1$, and is influenced by:

a) extraneous noise sources on the input and/or output,
b) other non-measurable inputs affecting the output,
c) the smoothing (window closing and window carpentry) procedures adopted in determining the estimate, and
d) the inherent transport (time) delays in the process.

Using equations (4.12), (4.16) and (4.20), it can be shown that the residual noise spectral density function $\hat{S}_{ZZ}(f)$ is related to the coherence function by:

$$\hat{S}_{ZZ}(f) = \hat{S}_{YY}(f)[1 - \hat{\gamma}^2(f)]$$                    ... (4.21)

## 4.5        TRANSFER FUNCTION CONFIDENCE LIMITS

The ordinary coherence function is also important for calculating the confidence limits associated with the transfer function modulus and phase estimates. The $100(1 - \alpha)\%$ confidence intervals for the transfer function modulus and phase estimates are respectively (Jenkins and Watts 1968)

$$|\hat{H}(f)| \{1 \pm \hat{r}(f)\}$$

$$\hat{\phi}(f) \pm \sin^{-1}\{\hat{r}(f)\}$$

where the parameter $\hat{r}(f)$ is the radius of the 'confidence' circle about the estimate $\hat{H}(f)$ of the process transfer function, and is given by

$$\hat{r}(f) = \sqrt{\frac{2}{\eta - 2} f_{2, \eta - 2}(1 - \alpha) \frac{1 - \hat{\gamma}^2(f)}{\hat{\gamma}^2(f)}}$$

in which $\eta$ is the number of degrees of freedom of the estimate, and $f_{2, \eta - 2}(1 - \alpha)$ is the sampling distribution function of the data. Note that the confidence intervals are small when the number of degrees of freedom $\eta$ is large, and when the coherence function $\hat{\gamma}^2(f)$ approaches unity.

## 4.6        TRANSFER FUNCTION ESTIMATION

We now investigate what happens when the transfer function estimates computed in practice from the input-output measurements are affected by:

1) extraneous or uncorrelated noise on either the input or output measurements, or both measurements;
2) data filtering of the input and output measurements;
3) data smoothing ('windowing') to minimise leakage;
4) time/transport delays.

### 4.6.1      Effect of Uncorrelated Noise

The more general case in which there is uncorrelated noise present in both the input and output measurements is illustrated in Figure 4.2. The Fourier transformed input and output measurements $\hat{U}(f)$ and $\hat{V}(f)$ consist of the 'true' signals $\hat{X}(f)$ and $\hat{Y}(f)$ and the extraneous noise components $\hat{M}(f)$ and $\hat{N}(f)$ such that

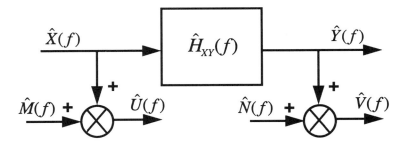

**Figure 4.2** Uncorrelated noise on the input and output measurements

$$\hat{U}(f) = \hat{X}(f) + \hat{M}(f)$$
$$\hat{V}(f) = \hat{Y}(f) + \hat{N}(f)$$

... (4.22)

Using the usual definitions, the corresponding spectral density estimates are:

$$\hat{S}_{UU}(f) = \hat{S}_{XX}(f) + \hat{S}_{MM}(f)$$
$$\hat{S}_{UV}(f) = \hat{S}_{XY}(f)$$
$$\hat{S}_{VV}(f) = \hat{S}_{YY}(f) + \hat{S}_{NN}(f)$$

... (4.23)

where it has been assumed that the true signals are uncorrelated with the noise signals (i.e. $\hat{S}_{XM}(f), \hat{S}_{YN}(f)$, etc. are zero), and the noise signals are uncorrelated $(\hat{S}_{MN}(f) = 0)$. Using equations (4.23), the measured transfer function $\hat{H}_{UV}(f)$ is related to the true transfer function $\hat{H}_{XY}(f)$ by

$$\hat{H}_{UV}(f) = \frac{\hat{S}_{UV}(f)}{\hat{S}_{UU}(f)} = \frac{\hat{H}_{XY}(f)}{1 + \dfrac{\hat{S}_{MM}(f)}{\hat{S}_{XX}(f)}}$$

... (4.24)

that is, it is modified by the noise to signal ratio on the *input* measurement only. Note that, as a consequence of the equivalence of the cross-spectral density functions (4.23), the *phase estimates* of $\hat{H}_{XY}(f)$ in (4.24) are uncontaminated by measurement noise on the input signal.

Similarly, the measured coherence function, $\hat{\gamma}_{UV}^2(f)$, and the true coherence function $\hat{\gamma}_{XY}^2(f)$ are related by

$$\hat{\gamma}_{UV}^2(f) = \frac{\left|\hat{S}_{UV}(f)\right|^2}{\hat{S}_{UU}(f)\,\hat{S}_{VV}(f)} = \frac{\hat{\gamma}_{XY}^2(f)}{\left(1 + \dfrac{\hat{S}_{MM}(f)}{\hat{S}_{XX}(f)}\right)\left(1 + \dfrac{\hat{S}_{NN}(f)}{\hat{S}_{YY}(f)}\right)}$$

... (4.25)

that is, it is modified by the noise to signal ratios on both the input and output measurements. It follows from equations (4.24) and (4.25) that the measurement noise should be as small as possible for reliable transfer function estimates.

### 4.6.2      Effect of Data Filtering

The data monitored on a process are normally sampled at equally spaced time intervals $\Delta t$ seconds apart, which corresponds to a sampling (or Nyquist) frequency $f_N = 1/(2\Delta t)$ Hz. To avoid bias in the correlation and spectral estimates due to 'aliasing', the frequencies above the Nyquist frequency must be removed (filtered) from the data prior to further analysis. Unless the filters on the input and output measurements have *identical transfer function characteristics*, the 'measured' and 'true' transfer functions will not be the same, as shown by the following analysis.

If the Fourier transformed input and output realizations after filtering are $\hat{C}(f)$ and $\hat{D}(f)$ respectively, then

$$\hat{C}(f) = F_{XC}(f)\,\hat{X}(f)$$
$$\hat{D}(f) = F_{YD}(f)\,\hat{Y}(f)$$

... (4.26)

The corresponding *filtered* input power and cross-spectral density estimates are given respectively by

$$\hat{S}_{CD}(f) = F_{XC}^{*}(f)\,F_{YD}(f)\,\hat{S}_{XY}(f)$$
$$\hat{S}_{CC}(f) = |F_{XC}(f)|^2\,\hat{S}_{XY}(f)$$

... (4.27)

Thus, the relationship between the filtered and 'true' transfer function is

$$\hat{H}_{CD}(f) = \left[\frac{F_{XC}^{*}(f)\,F_{YD}(f)}{|F_{XC}(f)|^2}\right]\hat{H}_{XY}(f)$$

... (4.28)

from which it follows that the true transfer function $\hat{H}_{XY}(f)$ is *not equivalent* to the filtered transfer function $\hat{H}_{CD}(f)$ unless the input and output filters have identical transfer function (gain/phase) characteristics; that is,

$$F_{XC}(f) = F_{YD}(f)$$

... (4.29)

Using the above spectral equations (4.27) and the corresponding relation for $\hat{S}_{DD}(f)$, it can be readily shown that the true and filtered coherence functions are equivalent; that is, they are unaffected by the filtering operation since $\hat{\gamma}_{CD}(f) = \hat{\gamma}_{XY}(f)$, at least in theory. In practice, however, the transfer function and coherence estimates *outside* the filter bandwidths are unreliable, and may give spurious results.

### 4.6.3 Effect of Data Smoothing (Windowing) to Minimise Leakage

Since the sampled data signals $\hat{x}(t)$ and $\hat{y}(t)$ are obtained for only a finite length of time ($0 \le t \le T$), the infinite Fourier transform is truncated and there is 'leakage' (loss) of information in the $T \to \infty$ range. The purpose of the smoothing (window) operation is to minimise this 'leakage' in information, which results in side lobes in the frequency domain spectral estimates.

#### 4.6.3.1 A Simulation Example

Its effects on the computed spectral functions may be illustrated by considering the Fourier transform of the periodic signal $x(t) = X_0 \sin(2\pi f' t)$. The theoretical autocorrelation and power spectral density functions (Appendix A4) are respectively:

$$R_{xx}(\tau) = \frac{X_0^2}{2} \cos(2\pi f' \tau) \qquad |\tau| < \tau_m$$

$$= 0 \qquad |\tau| \ge \tau_m$$

$$\text{and} \quad S_{xx}(f) = \frac{X_0^2 \tau_m}{2} \left[ \frac{\sin\{2\pi(f + f')\tau_m\}}{2\pi(f + f')\tau_m} + \frac{\sin\{2\pi(f - f')\tau_m\}}{2\pi(f - f')\tau_m} \right]$$

$$\text{... (4.30)}$$

The power spectral density function $S_{xx}(f)$ consists of individual sinc (or $\sin(x)/x$) functions centred on the resonant frequencies $\pm f'$, and an example of the power spectral density function $S_{xx}(f)$ computed from equations (4.30) is shown in Figure 4.3, in which $\tau_m = 20$ seconds, $f_N = 2$ Hz, $X_0^2 \tau_m / 2 = 1$, hence $\Delta t = 1/f_N = 0.25$ s and $\Delta f = 0.025$ Hz. Two cases are shown, one with an equal number of periods ($N = 20, f' = 1$ Hz) and one with an unequal number of periods ($N = 20.25, f' = 1.0125$ Hz). Note the single peak ordinate of correct magnitude in the former case and the reduced peak ordinate with leakage components in the latter case.

The leakage contributions of the side lobes on the resultant spectral estimates are noted as follows:

1) $S_{xx}(f)$ is zero when the sine terms are zero, that is, at frequencies $\Delta f = 1/(2\tau_m)$ apart. If the periodic signal is sampled every $\Delta t$ seconds and $M$ values of the correlation function are chosen such that $\tau_m (=M\Delta t)$

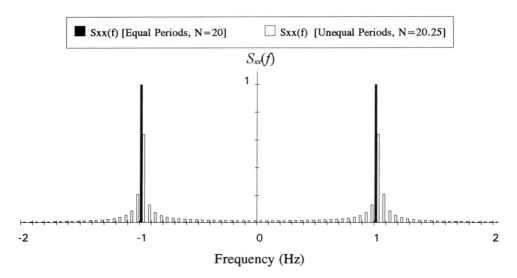

**Figure 4.3** Spectral density functions for a periodic signal showing leakage
components

is an integer number of periods ($\tau_m = NT'$, $T' = 1 / f'$), then the discrete
values of $S_{xx}(f)$ has the expected maxima $S_{xx}(f') = X_0^2 \tau_m / 2$ at
frequencies $f = \pm f'$ and zero values elsewhere ($|f_n| = n\Delta f$, $n=1,2,...$).
Thus there is *no leakage* due to the side lobes in the sine functions at
these frequencies, and the discrete estimates correspond to the
continuous spectral results. This point is emphasised by Brigham (1988)
and Burgess (1975), and is particularly relevant to the computation of
spectral estimates by fast Fourier transform (FFT) algorithms, which
have a fixed frequency interval $\Delta f = 1/(2\tau_m)$. However the spectral
estimates at intermediate frequencies still contain leakage components.

2)  If the maximum lag ($\tau_m$) in the correlation function or the length of
    record ($T$) for direct Fourier transform (FT) analysis are not equal to an
    integer number of periods, the side-lobe characteristics may cause
    considerable leakage in the spectral estimates at the desired frequencies
    ($f = \pm f'$ and $f_n = n\Delta f$, $n = 1,2,3,...$), thus resulting in discrepancies
    between the discrete and continuous spectral results. The two sine
    functions in equation (4.28) are 'out-of-phase' and interact to produce
    resultant maxima which are not coincident with the excitation frequency
    ($f'$). In the time domain there are discontinuities in the periodically
    extended sample signal as assumed by the FT computations (Burgess

1975), and there is a resultant mean value which gives rise to a leakage component at zero frequency.

### 4.6.3.2    The Windowing Operation

As we have seen (section 3.5, Chapter 3), a finite length variable is obtained from its infinite variable by weighting it with a *data window* function $w(t)$ such that,

$$
\begin{aligned}
\bar{y}(t,T) &= w(t)\,y(t), \quad 0 \le t \le T \\
&= 0, \qquad\qquad t \ge T
\end{aligned}
\qquad \text{... (4.31)}
$$

In a corresponding manner, the finite correlation functions are weighted by the '*lag*' *window* function $w(\tau)$ prior to Fourier transformation according to the equation

$$
\begin{aligned}
\overline{R}(\tau,\tau_m) &= w(\tau)\,R(\tau), \quad -\tau_m \le \tau \le \tau_m \\
&= 0, \qquad\qquad\quad \tau \ge |\tau_m|
\end{aligned}
\qquad \text{... (4.32)}
$$

Alternatively, the smoothed spectral estimators may be calculated from the convolution equation

$$
\overline{S}(f) = \int_{-\infty}^{\infty} W(f - f')\,S(f')\,df'
\qquad \text{... (4.33)}
$$

where the spectral window function $W(f)$ and the lag window $w(\tau)$ are Fourier transform pairs. Comparing equations (4.27) and (4.33), it can be seen that smoothing the spectral estimates by a spectral window $W(f)$ is not the same as smoothing by a filter function $F(f)$. Note that the window operation is a *convolution* in the frequency domain (or product in the time domain), whereas the filter operation is a *product* in the frequency domain (or convolution in the time domain).

### 4.6.3.3    Some Common Window Functions

In general, the time series data sampled on industrial processes contain broadband frequency components and the leakage reduction operation must simultaneously satisfy the conflicting requirements of statistical stability (i.e. minimise the magnitude of the side lobes) and frequency resolution. This conundrum has prompted extensive research into the design of 'optimum' window functions for real time digital signal processing, and a variety of methods (or properties) have been developed for selecting the 'optimum' window characteristics. Readers are referred to the comprehensive survey by Harris (1978) and standard texts (e.g. Oppenheim and Schafer (1989) and

Proakis and Manolakis (1992)) for further details.

The formulae for the rectangular ('box car'), Bartlett, Hanning and Parzen windows commonly used in practice are given in Table B1, Appendix B4. Plots of their respective *lag* and *spectral* window functions are shown in Figure 4.4, where $A = 2\pi f \tau_m$ and $\tau_m = 50$. We note in passing that the *data* windows are sometimes referred to as being *linear phase causal* since, on Fourier transformation, we obtain a generalised spectral window function of the form:

$$\int_0^T w(t)e^{-j\omega t}dt = W(f)e^{-j\omega T/2} \qquad \qquad ...(4.34)$$

where $W(f)$ is given by the respective spectral window function in Table B1, Appendix B, by replacing $\tau_m$ with $T/2$. This linear phase characteristic is due to the asymmetry of a signal $x(t)$ about the ordinate ($t = 0$ axis), and does not arise with *lag* windows because of their symmetry about the zero lag ($\tau = 0$) axis.

### 4.6.3.4    Spectral Window Properties and 'Optimisation' Procedures

Various criteria have been developed to define the key properties of the window functions, and these are either *statistically based* or *frequency response based* in their definition. In section 3.5.2, Chapter 3, we used the former approach to assess the accuracy of the spectral estimates from finite length records. Table B3.2, Appendix B, presents the key statistical properties for the above window functions using this method. Recall from equation (3.90) that the determination of an '*optimum*' window involves minimisation of both the *bias* and *variance* of the smoothed spectral estimates, that is, we need to make the mean square error function:

$$\overline{E^2}(f) = B^2(f) + \text{Var}[\hat{\overline{S}}(f)] \qquad \qquad ...(4.35)$$

as small as possible. One method was proposed in section 3.5.2, Chapter 3. Jenkins and Watts (1968) suggest a practical *window closing* technique in which the resolution bandwidth is narrowed by increasing the record length ($T$) or the maximum number of lags ($\tau_m$) until window instabilities are detected in the smoothed spectral estimates. (The resolution bandwidth may also be adjusted by selecting a different window function from Table B3.1, Appendix B3, but this technique (termed *window carpentry*) is usually of second order importance to the window closing technique.) Ideally, we would like to minimise equation (4.35) to obtain an optimum record length ($T_{opt}$) or maximum lags ($\tau_{opt}$), but this requires a knowledge of the true spectral density function $S(f)$ which is usually not available in practice.

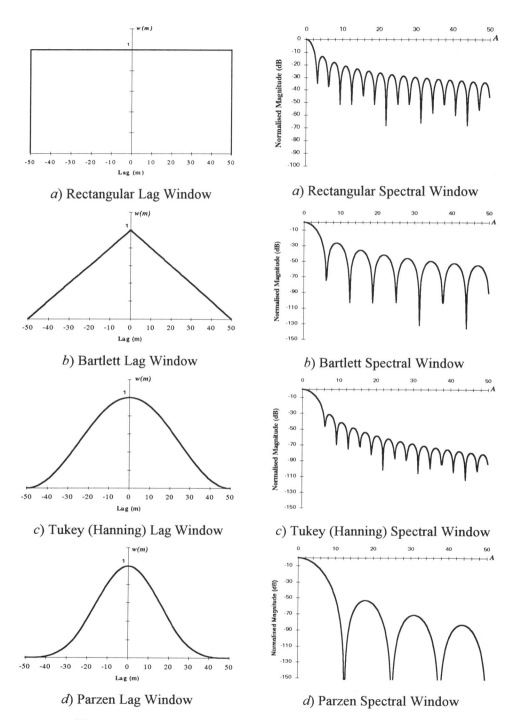

a) Rectangular Lag Window

a) Rectangular Spectral Window

b) Bartlett Lag Window

b) Bartlett Spectral Window

c) Tukey (Hanning) Lag Window

c) Tukey (Hanning) Spectral Window

d) Parzen Lag Window

d) Parzen Spectral Window

**Figure 4.4** Plots of some lag and spectral window functions

Approximate expressions are derived by Jenkins and Watts (1968) as an aid in selecting a window with properties appropriate for an application, and these form the basis of the window properties given in Table B3.2, Appendix B. However, these statistically based expressions only provide a subjective guide to selecting the 'optimum' window characteristics, and a more optimal design method based was developed by Kaiser as outlined in the following section (after Oppenheim and Schafer 1989).

### 4.6.3.5    The Kaiser Window Design Method

Kaiser found that a zeroth-order Bessel function of the first kind yielded a near-optimal *data* window of the form:

$$w_K(t) = \frac{I_0[\beta(1 - \{(t - \alpha)/\alpha\}^2)^{1/2}]}{I_0(\beta)}; \quad 0 \le t \le T \qquad \text{...(4.36a)}$$

$$= 0; \qquad\qquad\qquad\qquad t > T$$

where

$$\alpha = T/2$$

$$\beta = \begin{cases} 0.1102(B - 8.7) & B > 50 \\ 0.5842(B - 21)^{0.4} + 0.07886(B - 21) & 21 \le B \le 50 \\ 0.0 & B < 21 \end{cases} \quad \text{... (4.36b)}$$

$$B = -20 \log_{10} \delta$$

and the parameter $\delta$ is the *peak approximation error* between the frequency response of an 'ideal' filter and the 'ideal' filter convolved with a spectral window function, equation (4.33) and Figure 4.5. For a wide range of conditions, the value of $\delta$ (a measure of the bias) is determined by the choice of $\beta$, and one obtains the empirical equations (4.36b). For example, a *rectangular window* has values of $\beta = 0$ and $B = 2$, which correspond to a peak approximation error $(\delta)$ of $\pm 9\%$. For a fixed $\delta$, the bandwidth $\Delta\omega$ in the transition region is defined as

$$\Delta\omega = \omega_s - \omega_p \qquad \text{... (4.37)}$$

where $\omega_p$ is the passband cutoff frequency of the lowpass filter such that $|\overline{H}(\omega)| \ge |H(\omega)| - \delta$, and $\omega_s$ is the stopband cutoff frequency such that $|\overline{H}(\omega)| \le \delta$. To achieve prescribed values of $B$ and $\Delta\omega$, The number of samples $N (T = N\Delta t)$ for the data record must satisfy

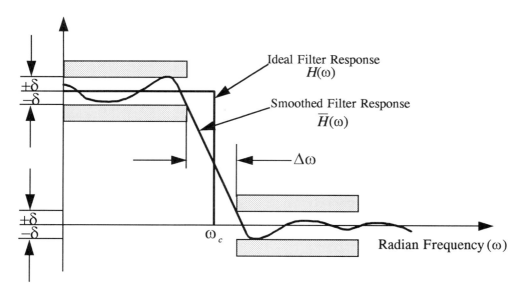

**Figure 4.5** Illustration of 'ideal' and smoothed filter responses

$$N = \frac{B - 8}{2.285\Delta\omega} \qquad \text{... (4.38)}$$

Equation (4.38) predicts the values for $N$ to within $\pm2$ samples over a wide range of values for $B$ and $\Delta\omega$. A major advantage of the Kaiser window design method is that, unlike the 'statistical' window design methods described previously, it requires almost no iteration or trial and error. There are equivalence relationships between the Kaiser windows and those discussed in the previous section, and comparisons with their equivalent Kaiser spectral window properties are given in Table B4.3, Appendix B4.

### 4.6.4 Time/Transport Delay Compensation

Significant bias can also occur in the estimation of cross-spectra if there is a large transport delay in the transfer of information from the input to the output. This bias is reduced by *alignment* of the input and output time series. A property of Fourier transforms is that a shift in the time domain is equivalent to a phase shift in the frequency domain. Hence if the output $\hat{y}(t)$ is shifted by the transport delay $(\tau_d)$ to align it with the input $\hat{x}(t)$, then the resultant Fourier transform is

$$\int_0^T \hat{y}(t - \tau_d)\, e^{-j2\pi ft}\, dt = \hat{Y}(f)\, e^{-j2\pi f\tau_d} \qquad \text{... (4.39)}$$

Equivalent results are obtained by aligning the $\tau = 0$ axis with the maximum peak in the cross-correlation function $\hat{R}_{xy}(\tau)$, which would occur at $\tau = \tau_d$. In this case the resultant Fourier transform is

$$\int_{-\infty}^{\infty} \hat{R}_{xy}(\tau - \tau_d) e^{-2\pi ft} \, d\tau = \hat{S}_{XY}(f) e^{-2\pi f\tau_d} \qquad \dots (4.40)$$

If the cross-spectral function is approximately constant over the frequency bandwidth of the spectral window, and the input and output power spectral density functions are assumed to be unbiased, then the bias in the coherence function estimates is given by (Jenkins and Watts 1968):

$$\hat{B}(f) = E\left[\hat{\gamma}^2_{XY}(f) - \gamma^2_{XY}(f)\right]$$

$$= -\frac{C_w}{\tau_m^2} \gamma^2_{XY}(f) \left[\frac{d\phi_{XY}(f)}{df}\right]^2$$

where $\gamma^2_{XY}(f)$ is the theoretical coherence function, $\phi_{XY}(f)$ is the theoretical phase angle, $\tau_m$ is the maximum lag and $C_w$ is a coefficient dependent on the spectral window $W(f)$.

If $\gamma^2_{XY}(f) \approx 1$ and $\phi_{XY}(f) \approx -2\pi f\tau_d$ then

$$\hat{B}(f) \approx -4\pi^2 C_w \left[\frac{\tau_d}{\tau_m}\right]^2$$

and the mean coherence estimate is

$$\hat{\gamma}^2(f) = E\left[\hat{\gamma}^2_{XY}(f)\right] = 1 - 4\pi^2 C_w \left[\frac{\tau_d}{\tau_m}\right]^2 \qquad \dots (4.41)$$

As equation (4.41) shows, alignment of the cross-correlation function ($\tau_d = 0$) results in a mean coherence estimate of unity, which is the same as the theoretical coherence value $\gamma^2_{XY}(f)$. The effects of transport delays on the coherence estimates have been investigated by Romberg and Harris (1978), and are demonstrated in the simulation example discussed in the next section.

## 4.7     COMPUTATIONAL PROCEDURES AND EXAMPLE

The methods for computing the various correlation, spectral density, transfer function and coherence estimates discussed above are shown schematically in Figure 4.6. Both methods are applied to the analysis of a second order discrete process in the demonstration programs supplied with this text.

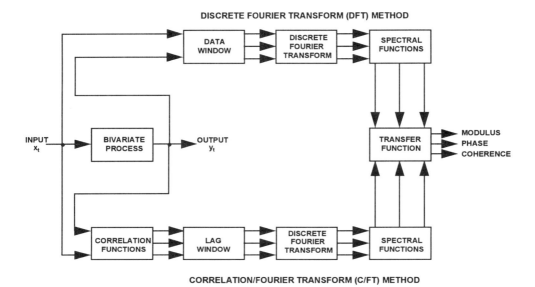

**Figure 4.6** Schematic diagram of spectral analysis methods

### 4.7.1     Discrete Fourier Transform (DFT) Method

In the first method, termed the *discrete Fourier transform (DFT)* method, the digitised input and output measurements are Fourier transformed using the equations:

$$\hat{\overline{X}}(f_k) = \Delta t \sum_{n=0}^{N-1} \hat{x}(t_n) w(t_n) e^{-j2\pi kn/N}$$

$$\hat{\overline{Y}}(f_k) = \Delta t \sum_{n=0}^{N-1} \hat{y}(t_n) w(t_n) e^{-j2\pi kn/N}$$

... (4.42)

where $k = 0, 1, \ldots , (N–1)$, $x(t_n)$ and $y(t_n), 0 < t_n < T$, are sampled at $N$ equally spaced points at a time interval $\Delta t$ apart, the total sample length $T = N\Delta t$, $w(t_n)$ is the data window function and $f_k = k / T = k / N\Delta t$ is the $k$th frequency value of the transformed variables. Note that the smoothed frequency domain functions $\overline{X}(f_k)$ and $\overline{Y}(f_k)$ are only unique up to the Nyquist frequency, $f_N = 1 / (2\Delta t)$, and that frequencies greater than $f_N$ must be filtered out to avoid the aliasing problems discussed in Chapter 3.

The *one-sided* spectral density estimates are then calculated from the equations:

$$\hat{G}_{XX}(f_k) = \frac{2}{N\Delta t}\left|\hat{X}(f_k)\right|^2$$

$$\hat{G}_{YY}(f_k) = \frac{2}{N\Delta t}\left|\hat{Y}(f_k)\right|^2 \qquad k = 0, 1, \dots, N/2 \quad \dots (4.43)$$

$$\hat{G}_{XY}(f_k) = \frac{2}{N\Delta t}\left[\hat{X}^*(f_k)\hat{Y}(f_k)\right]$$

It should be noted that $\hat{G}_{XX}(f_k)$, $\hat{G}_{XY}(f_k)$ and $\hat{G}_{YY}(f_k)$ in equations (4.43) are defined for *positive* frequencies ($0 \le f_k \le f_N$) only, and are related to the corresponding *two-sided* spectral density estimates $\hat{S}_{XX}(f_k)$, $\hat{S}_{XY}(f_k)$ and $\hat{S}_{YY}(f_k)$, $-f_N \le f_k \le f_N$, by 'folding' them about the zero frequency axis and summing the spectral density estimates at each frequency $f_k$.

The spectral density estimates in equations (4.43) can be calculated using either a standard discrete Fourier transform (DFT) or fast Fourier transform (FFT) algorithms (Brigham 1988).

### 4.7.2    Correlation/Fourier Transform (C/FT) Method

The second method, termed the correlation/Fourier transform (C/FT) method, computes the spectral density estimates from their corresponding correlation functions using the discrete convolution equations (cf. equations (3.79)):

$$\hat{R}_{xx}(m) = \frac{1}{N-|m|}\sum_{n=0}^{N-|m|-1}\hat{x}_n\hat{x}_{n+m}$$

$$\hat{R}_{yy}(m) = \frac{1}{N-|m|}\sum_{n=0}^{N-|m|-1}\hat{y}_n\hat{y}_{n+m} \qquad \dots (4.44)$$

$$\hat{R}_{xy}(m) = \frac{1}{N-|m|}\sum_{n=0}^{N-|m|-1}\hat{x}_n\hat{y}_{n+m}$$

where the time lag $\tau_m = m\Delta t$ ($|m| = 0,1,\dots,M$) and $M$ is the maximum number of lags. The smoothed (windowed) spectral density estimates are given by:

$$\hat{\bar{G}}_{XX}(f_n) = 2\Delta t\left[\hat{R}_{xx}(0)+2\sum_{m=1}^{M-1}w(m)\hat{R}_{xx}(m)\cos(2\pi km/M)\right]$$

$$\hat{\bar{G}}_{YY}(f_n) = 2\Delta t\left[\hat{R}_{yy}(0)+2\sum_{m=1}^{M-1}w(m)\hat{R}_{yy}(m)\cos(2\pi km/M)\right] \qquad \dots (4.45)$$

$$\hat{\bar{G}}_{XY}(f_n) = 2\Delta t\left[\hat{R}_{xy}(0)+2\sum_{m=0}^{M-1}w(m)\{\hat{R}_{xy}(m)e^{-j2\pi km/M} + \hat{R}_{yx}(m)e^{j2\pi km/M}\}\right]$$

where $w(m)$ is the discrete *lag window* function and $n = 0, 1, \ldots, M/2$.

The transfer function and coherence estimates are computed from equations (4.43) and (4.45) using the equations (4.16) and (4.20) respectively (see also Appendix A4).

### 4.7.3    A Simulation Example

The above theoretical and computational concepts are demonstrated with a simulation example. Consider a second order continuous process with a transfer function of the form (in Laplace transform notation):

$$H(s) = \frac{Y(s)}{X(s)} = \frac{Ke^{-s\tau_d}}{(s^2 + 2\xi\omega_n s + \omega_n^2)} \qquad \ldots (4.46)$$

where $K$ is the static gain, $\xi$ is the damping ratio, $\omega_n$ is the natural frequency (rad/s), $\tau_d$ is the process transport delay and $s$ is the Laplace operator ($s = \sigma + j\omega$). Assuming that the process has complex roots, the time constants are:

$$T_1 = \frac{e^{j\lambda}}{\omega_n}; \quad T_2 = \frac{e^{-j\lambda}}{\omega_n} \qquad \ldots (4.47)$$

The equivalent discrete process transfer function is given by (see Chapter 3):

$$H(z) = \frac{Y(z)}{X(z)} = \frac{(\beta_1 z^{-1} + \beta_2 z^{-2})z^{-d}}{(\alpha_0 + \alpha_1 z^{-1} + \alpha_2 z^{-2})} \qquad \ldots (4.48)$$

which is equivalent to the discrete [ARMA(2,2)] process difference equation:

$$\begin{aligned} y_t &= -\alpha_1 y_{t-1} - \alpha_2 y_{t-2} + \beta_1 x_{t-1-d} + \beta_2 x_{t-2-d}; \quad t > d+1 \\ y_t &= 0; \qquad\qquad\qquad\qquad\qquad\qquad\qquad\qquad t \le d+1 \end{aligned} \qquad \ldots (4.49)$$

and the discrete parameters are given by:

$$\begin{aligned} &\alpha_1 = -(S_1 + S_2); \text{ where } S_1 = e^{-\Delta/T_1}, S_2 = e^{-\Delta/T_2} \\ &\alpha_2 = S_1 S_2 \\ &\beta_1 = \frac{K}{T_1 - T_2}[T_1(1 - S_1) - T_2(1 - S_2)] \\ &\beta_2 = \frac{-K}{T_1 - T_2}[T_1(1 - S_1)S_2 - T_2(1 - S_2)S_1] \end{aligned} \qquad \ldots (4.50)$$

The simulation tests discussed below were performed using the computer program SISO#SIM.EXE supplied with the text for IBM/PC compatible computers operating under Windows 3.1 and higher (see Appendix C3 for installation and operational instructions).

The *aim* of this simulation is to demonstrate the effects of transport delays on the computed coherence estimates (cf.. Romberg and Harris 1978). The simulation tests were performed using the following computational procedure:

1) $M$ data values of the input, $x(t)$, are generated using a random number generator and passed through a digital filter to eliminate *aliasing*;
2) $M$ data values of the output, $y(t)$, are generated using equation (4.49);
3) the smoothed spectral density estimates are computed using equations (4.43) and (4.45) and a Tukey (Hanning) data/lag window function;
4) steps 1 to 3 are repeated for $L$ blocks of data until the total number of data points $N = L \times M$ is reached.
5) the block averaged spectral density estimates are calculated from the equation:

$$\tilde{G}(f_m) = \frac{1}{L}\sum_{l=1}^{L}\overline{G}_l(f_m) \qquad m = 0, 1, \ldots, M/2$$

The graphical charts generated by the programs display the development of the various functions at each step in the calculation loop. The final charts and print log present the final results. Steps 1 to 5 are repeated for new delay values, and the effects on the coherence estimates for the various delay ratios $(d/M)$ are compared in Figure 4.7, where the theoretical curve is given by

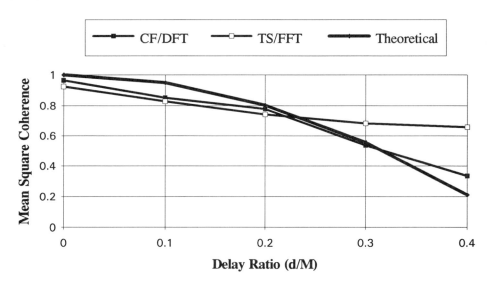

**Figure 4.7** Effects of time delays on mean square coherence estimates

equation (4.41). Note how the time series/discrete fast Fourier transform (TS/FFT) method performs better than the correlation (CF/DFT) method at the higher delay ratios. This is due to the influence of the time delays associated with the data window, equation (4.34), which tend to 'obscure' the dependence on delay ratio at the higher delay ratios. As discussed above, *alignment* of the time series or, equivalently, the cross-correlation function minimise the effects of inherent time delays in a process.

## REFERENCES

Brigham, E. O., 1988, *The Fast Fourier Transform and its Applications*, Prentice-Hall, Englewood Cliffs, NJ.

Burgess, J. C., 1975, On Digital Spectrum Analysis of Periodic Signals, *J. Acoustical Society of America* **58**(3), 556-567.

Harris, F. J., 1978, On the Use of Windows for Harmonic Analysis with the Discrete Fourier Transform, *Proc. IEEE*, **66**, pp 51-83, January.

Jenkins, G. M., & Watts, D. G., 1968, *Spectral Analysis and its Applications*, Holden-Day, San Francisco.

Oppenheim, A. B., & Schafer, R. W., 1989, *Discrete-Time Signal Processing*, Prentice-Hall, Englewood Cliffs, NJ.

Priestly, M. B., 1981, Estimation of Transfer Functions in Closed Loop, *Automatica*, **5**, 623-632.

Proakis, J. G., & Manolakis, D. G., 1992, *Digital Signal Processing: Principles, Algorithms and Applications*, Macmillan, New York.

Romberg, T. M., & Harris, R. W., 1978, A note on the Spectral Identification of Systems with Inherent Transport Delays, *Proc. Inst. of Radio and Electronic Engineers Australia*, **39**(6), 97-103.

## TUTORIAL EXERCISES

1) Choose one of the processes given in Figure 4.8.
   a) Show that it can be modelled by an ordinary differential equation of the form:

$$a_2 \frac{d^2 y(t)}{dt^2} + a_1 \frac{dy(t)}{dt} + a_0 y(t) = X(x,t)$$

   where $X(x,t)$ represents, respectively, the dynamic load on the beam (Figure 4.8*a*) or suspension system (Figure 4.8*b*), the source voltage (Figure 4.8*c*) and the change in level in reservoir 1 of the reservoir system (Figure 4.8*d*).
   b) Rewrite this equation in the differential operator form

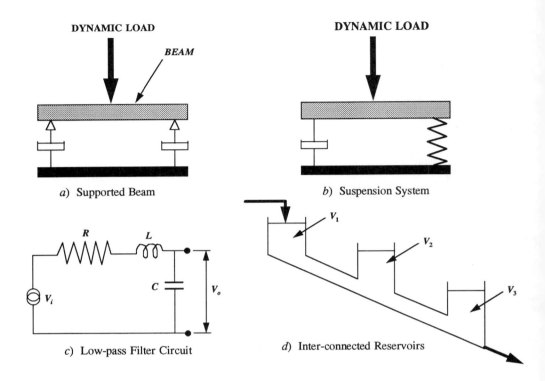

**Figure 4.8**  Tutorial examples of second order processes

$$(1 + T_1 D)(1 + T_2 D)y(t) = X(x,t)$$

and define the process time constants $(T_1, T_2)$ in terms of their physical parameters.

c) Derive the continuous system transfer function $G(f)$ in modulus and phase.

d) For a pulsed input, show that the equivalent discrete system transfer function $G(z)$ is given by an equation of the form

$$G(z) = \frac{\beta_0 + \beta_1 z^{-1} + \beta_2 z^{-2}}{(1 - S_1 z^{-1})(1 - S_2 z^{-1})}$$

e) Derive the discrete system transfer function $G_D(f)$ in modulus and phase.

f) Choose realistic values for the process parameters and draw plots comparing either the time responses or the frequency responses of the continuous and discrete processes for either a unit pulse input at $t = 0$ $[x(t) = \delta(t - t_0), t_0 = 0]$ or a unit step input $[x(t) = 1]$.

**HINT**: *Supported Beam and Suspension System* – The process outputs in these cases are the beam/suspension system deflections. Sum the forces acting on the beam in each case, and note that the beam stiffness $k = 48EI/L^3$, where $E$ is the Young's modulus, $I$ is the moment of inertia of the beam and $L$ is the span between the supports.

**HINT**: *Inter-connected Reservoirs* – Obtain the relationship between the process input (reservoir 1 volume, $V_1$) and the output (reservoir 3 volume, $V_3$) from volume balances on reservoirs 2 and 3. Note that the steady state flows into reservoir 1 and out of reservoir 3 are equal, but the inlet flow surges as stated in (1f) above and increases reservoir 1 volume ($V_1$).

2) Using the time constants calculated in (1f) above, compute the natural frequency ($\omega_n$) and damping ratio ($\xi$) of your selected process. Use the SISO#SIM.EXE program to investigate the time and frequency reponse characteristics of the process to random fluctuations in the input, $X(x,t)$. What observations can you make about the changes in your results with variations in one or more of the following analysis parameters:
   a) a range of small to large number of (data) Points/Block;
   b) a range of low to high number of Total (data) Points;
   c) a range of low to high values of Time Interval between samples (but noting *aliasing* requirements); and/or
   d) a range of low to high values of output noise/input signal (Z/X) Variance Ratios?

   NOTE: The results of SISO#SIM.EXE calculations are printed to a Print Log, which can be accessed via the Window/Open Print Log menu option. The Print Log results and the various plots generated by the program may be printed directly using the File/Print menu option. The Print Log results can also be copied to other plot programs such as MS Excel using the Windows Clipboard facility.

3) Using the SISO#SIM.EXE program, repeat the simulation example described above and so obtain the results presented in Figure 4.7. (See the NOTE in exercise (2) above for printing and plotting your results.)

4) Using equations (4.12), (4.16) and (4.20), prove that

$$\hat{S}_{ZZ}(f) = \hat{S}_{YY}(f)[1 - \hat{\gamma}^2(f)]$$

as stated in equation (4.21). What is the significance of this equation?

# APPENDIX A4

**Table A4.1** Summary of the basic relations for stationary processes

| FUNCTION | DEFINITION |
|---|---|
| Cumulative Probability Distribution | $P(x) = \text{Prob}\ \{x < x(t) < x + \Delta x\} = \int\limits_{x}^{x+\Delta x} p(x)\,dx$ |
| Probability Density Function | $p(x) = \dfrac{dP(x)}{dx};\ \ p(x,y) = \dfrac{\partial}{\partial y}\left(\dfrac{\partial P(x,y)}{\partial x}\right)$ |
| Expected Value Operator | $E[g(x)] = \int\limits_{0}^{1} g(x)\,dP(x) = \int\limits_{-\infty}^{\infty} g(x)\,p(x)\,dx$ |
| Mean Value (1st Moment) | $\bar{x} = E[x] = \int\limits_{-\infty}^{\infty} x\,p(x)\,dx = \lim\limits_{T \to \infty} \dfrac{1}{T} \int\limits_{0}^{T} x(t)\,dt$ |
| Mean Square (2nd Moment) | $\overline{x^2} = E[x^2] = \int\limits_{-\infty}^{\infty} x^2 p(x)\,dx = \lim\limits_{T \to \infty} \dfrac{1}{T} \int\limits_{0}^{T} x^2(t)\,dt$ |
| Variance | $\sigma_x^2 = E[(x-\bar{x})^2] = \lim\limits_{T \to \infty} \dfrac{1}{T} \int\limits_{0}^{T} \{x(t) - \bar{x}\}^2\,dt$ |
| Convolution Integral | $y(t) = \int\limits_{0}^{\infty} h(\tau)x(t-\tau)\,d\tau;\ \ \ h(\tau) = 0,\ \ \tau < 0$ |
| Input Autocorrelation | $R_{xx}(\tau) = E[x(t)\,x(t+\tau)] = \lim\limits_{T \to \infty} \dfrac{1}{T} \int\limits_{0}^{T} x(t)\,x(t+\tau)\,dt$ |
| Output Autocorrelation | $R_{yy}(\tau) = E[y(t)\,y(t+\tau)] = \lim\limits_{T \to \infty} \dfrac{1}{T} \int\limits_{0}^{T} y(t)\,y(t+\tau)\,dt$ |
| Input-Output Crosscorrelation | $R_{xy}(\tau) = E[x(t)\,y(t+\tau)] = \lim\limits_{T \to \infty} \dfrac{1}{T} \int\limits_{0}^{T} x(t)\,y(t+\tau)\,dt$ |

**Table A4.1** (Cont)

| FUNCTION | DEFINITION |
|---|---|
| Input Autocovariance | $C_{xx}(\tau) = E[\{x(t) - \bar{x}\}\{x(t+\tau) - \bar{x}\}] = R_{xx}(\tau) - \bar{x}^2$ |
| Output Auto-covariance | $C_{yy}(\tau) = E[\{y(t) - \bar{y}\}\{y(t+\tau) - \bar{y}\}] = R_{yy}(\tau) - \bar{y}^2$ |
| Input-Output Cross-covariance | $C_{xy}(\tau) = E[\{x(t) - \bar{x}\}\{y(t+\tau) - \bar{y}\}] = R_{xy}(\tau) - \overline{xy}$ |
| Weiner-Hopf Con-volution Integral | $R_{xy}(\tau) = \int_0^\infty R_{xx}(\tau - \lambda) h(\lambda) \, d\lambda$ |
| Correlation Function Properties | $R_{xx}(\tau) = R_{xx}(-\tau); \ R_{yy}(\tau) = R_{yy}(-\tau); \ R_{xy}(\tau) = R_{yx}(-\tau)$ |
| Input Power Spectral Density | $S_{XX}(f) = \underset{T \to \infty}{\mathrm{Lim}} \frac{1}{T} \{X^*(f) X(f)\} = 2\int_0^\infty R_{xx}(\tau) e^{-j2\pi f\tau} \, d\tau$ |
| Output Power Spectral Density | $S_{YY}(f) = \underset{T \to \infty}{\mathrm{Lim}} \frac{1}{T} \{Y^*(f) Y(f)\} = 2\int_0^\infty R_{yy}(\tau) e^{-j2\pi f\tau} \, d\tau$ |
| Input-Output Cross-spectral Density | $S_{XY}(f) = \underset{T \to \infty}{\mathrm{Lim}} \frac{1}{T} \{X^*(f) Y(f)\} = \int_{-\infty}^\infty R_{xy}(\tau) e^{-j2\pi f\tau} \, d\tau$ |
| Co-spectral Density | $C_{XY}(f) = \Re[S_{XY}(f)] = \int_{-\infty}^\infty R_{xy}(\tau) \cos(2\pi f\tau) \, d\tau$ |
| Quad-spectral Density | $Q_{XY}(f) = \Im[S_{XY}(f)] = \int_{-\infty}^\infty R_{xy}(\tau) \sin(2\pi f\tau) \, d\tau$ |
| Spectral Density Properties | $S_{XX}(f) = S_{XX}^*(f); \ S_{YY}(f) = S_{YY}^*(f); \ S_{XY}(f) = S_{YX}^*(f)$ |
| Transfer Function Modulus & Phase | $|H(f)| = \dfrac{|S_{XY}(f)|}{S_{XX}(f)}; \quad \phi(f) = \arctan\left(\dfrac{-Q_{XY}(f)}{C_{XY}(f)}\right)$ |

# APPENDIX B4

**Table B4.1** Some common data, lag and spectral window functions

| Window | Data Window $w(t)$ $[0 \le t \le T]$ | Lag Window $w(\tau)$ $[-\tau_m \le \tau \le \tau_m]$ | Spectral Window $W(f)$ $[-\infty < f < \infty]$ |
|---|---|---|---|
| Rectan-gular | $= 1 \quad t < T$ <br> $= 0 \quad t \ge T$ | $= 1 \quad \lvert \tau \rvert < \tau_m$ <br> $= 0 \quad \lvert \tau \rvert \ge \tau_m$ | $= 2\tau_m \left\{ \dfrac{\sin(2\pi f \tau_m)}{2\pi f \tau_m} \right\}$ |
| Bartlett | $= \dfrac{2t}{T} \quad 0 \le t \le T/2$ <br> $= 2\left(1 - \dfrac{t}{T}\right)$ <br> $T/2 < t \le T$ <br> $= 0 \quad t > T$ | $= 1 - \dfrac{\lvert \tau \rvert}{\tau_m} \quad \lvert \tau \rvert < \tau_m$ <br> $= 0 \quad \lvert \tau \rvert \ge \tau_m$ | $= \tau_m \left\{ \dfrac{\sin(\pi f \tau_m)}{\pi f \tau_m} \right\}^2$ |
| Hanning | $= \dfrac{1}{2}\left\{1 - \cos\left(\dfrac{2\pi t}{T}\right)\right\}$ <br> $t \le T$ <br> $= 0 \quad t > T$ | $= \dfrac{1}{2}\left\{1 + \cos\left(\dfrac{\pi \tau}{\tau_m}\right)\right\}$ <br> $\lvert \tau \rvert < \tau_m$ <br> $= 0 \quad \lvert \tau \rvert \ge \tau_m$ | $= \tau_m \left\{ \dfrac{\sin(2\pi f \tau_m)}{2\pi f \tau_m} \right\}$ <br> $\left\{ \dfrac{1}{1 - (2 f \tau_m)^2} \right\}$ |
| Parzen | $= 1 - 6\left(\dfrac{2t - T}{T}\right)^2$ <br> $+ 6\left(\dfrac{\lvert 2t - T \rvert}{T}\right)^3$ <br> $\dfrac{T}{4} \le t \le \dfrac{3T}{4}$ <br> $= 2\left(1 - \dfrac{\lvert 2t - T \rvert}{T}\right)^3$ <br> $\dfrac{3T}{4} < t < \dfrac{T}{4}$ <br> $= 0 \quad t > T$ | $= 1 - 6\left(\dfrac{\tau}{\tau_m}\right)^2$ <br> $+ 6\left(\dfrac{\lvert \tau \rvert}{\tau_m}\right)^3 \quad \lvert \tau \rvert \le \dfrac{\tau_m}{2}$ <br> $= 2\left(1 - \dfrac{\lvert \tau \rvert}{\tau_m}\right)^3$ <br> $\dfrac{\tau_m}{2} < \lvert \tau \rvert \le \lvert \tau_m \rvert$ <br> $= 0 \quad \lvert \tau \rvert > \tau_m$ | $= \dfrac{3}{4}\tau_m \left\{ \dfrac{\sin(\pi f \tau_m / 2)}{\pi f \tau_m / 2} \right\}^4$ |

**Table B4.2** Statistical properties of some commonly used windows

| Spectral Window | Approximate Bias $B(f)$ | Variance Ratio | $\chi^2$ Degrees of Freedom | Resolution Bandwidth |
|---|---|---|---|---|
| Rectangular | $0$ | $2\dfrac{\tau_m}{T}$ | $\dfrac{T}{\tau_m}$ | $\Delta f_R = \dfrac{1}{T}$ $= \dfrac{1}{2\tau_m}$ |
| Bartlett | $\dfrac{1}{\tau_m}\displaystyle\int_{-\infty}^{\infty} -\lvert\tau\rvert R_{xx}(\tau)e^{-j2\pi f\tau}\,d\tau$ | $\dfrac{2\tau_m}{3T}$ | $3\dfrac{T}{\tau_m}$ | $\Delta f_R = \dfrac{3}{T} = \dfrac{3}{2\tau_m}$ |
| Hanning | $\dfrac{\pi^2}{4\tau_m^2}\displaystyle\int_{-\infty}^{\infty} -\tau^2 R_{xx}(\tau)e^{-j2\pi f\tau}\,d\tau$ $+O\!\left(\dfrac{1}{M^4}\right)$ | $\dfrac{3\tau_m}{4T}$ | $2.667\dfrac{T}{\tau_m}$ | $\Delta f_R = \dfrac{8}{3T} = \dfrac{4}{3\tau}$ |
| Parzen | $\dfrac{6}{\tau_m^2}\displaystyle\int_{-\infty}^{\infty} -\tau^2 R_{xx}(\tau)e^{-j2\pi f\tau}\,d\tau$ $+O\!\left(\dfrac{1}{M^3}\right)$ | $0.539\dfrac{\tau_m}{T}$ | $3.71\dfrac{T}{\tau_m}$ | $\Delta f_R = \dfrac{3.72}{T}$ $= \dfrac{1.86}{\tau_m}$ |

Bias (cf. equation (3.90)): $\quad B(f) \approx \displaystyle\int_{-\infty}^{\infty} \{w(\tau) - 1\} R(\tau)e^{-j2\pi f\tau}\,d\tau$

$\sigma^2;\sigma^2$ ratio: $\quad \mathrm{Var}[\hat{\bar{S}}(f)] \approx \dfrac{S''(f)}{T}\displaystyle\int_{-\infty}^{\infty} w^2(\tau)\,d\tau\,; \quad \dfrac{\mathrm{Var}[\hat{\bar{S}}(f)]}{\mathrm{Var}[S(f)]} \approx \dfrac{1}{T}\displaystyle\int_{-\infty}^{\infty} w^2(\tau)\,d\tau = \dfrac{I}{T}$

where $S''(f) \approx \mathrm{Var}[S(f)]$ is the second differential with respect to $f$ of the true power spectral density $S(f)$, and $I = \displaystyle\int_{-\infty}^{\infty} w^2(\tau)\,d\tau$.

Resolution Bandwidth: $\quad \Delta f_r = \dfrac{1}{\displaystyle\int_{-\infty}^{\infty} w^2(\tau)\,d\tau} = \dfrac{1}{I}$

$\chi^2$ Degrees of Freedom: $\quad v \approx \dfrac{2T}{\displaystyle\int_{-\infty}^{\infty} w^2(\tau)\,d\tau} = \dfrac{2T}{I}$

**Table B4.3** Frequency response properties of some common windows
and their equivalent Kaiser spectral window*

| Window Type | Peak Sidelobe Relative Amplitude | Approx. Width of Mainlobe (rad/s) | Peak Approx'n Error $20\log_{10}\delta$ | Equivalent Kaiser Spectral Window β Value (4.36b) | Equivalent Kaiser Spectral Window Transition Width $\Delta\omega$, (4.38) |
|---|---|---|---|---|---|
| Rectan-gular | −13 dB | $4\pi/(N+1)$ | −21 dB | 0 | $1.81\pi/N$ |
| Bartlett | −25 dB | $8\pi/N$ | −25 dB | 1.33 | $2.37\pi/N$ |
| Hanning | −31 dB | $8\pi/N$ | −44 dB | 3.86 | $5.01\pi/N$ |
| Parzen** | −53 dB | $1\pi/N$ | −65 dB | 6.02 | $7.94\pi/N$ |

\*     After Oppenheim and Schafer (1989).

\*\*   Estimates based on Spectral Window frequency response (Figure 4.4$d$).

# APPENDIX C4

## SISO#SIM PROGRAM INSTALLATION AND OPERATION

SISO#SIM.EXE computes the correlation, spectral, transfer and ordinary coherence functions for a second order process and a range of output noise/input signal ratios. It has been designed to run under Windows 3.1 with video display settings of 640×480 pixels×256 colours, and also runs under Windows 95 with the same video display settings. Refer to the **readme.txt** document for further information.

## C4.1    INSTALLATION

1) Insert disc in the FDD and type **a:\setup filename** in the Windows 3.1 Program Manager File/RUN menu.
2) The **setup.bat** file copies *all* the files on the floppy disc to the hard drive directory **c:\filename**. (NOTE: The setup procedure will terminate if **filename** is not specified.)
3) The **setup.bat** file also decompresses the RUN library files **cable.dll** and **rlzrun20.rts** and instals them into a newly created Windows sub-directory **c:\windows\rlzrun20**.
4) For ease of operation, it is recommended that the four simulation programs in the **c:\filename** directory be installed as a program group icon with their separate program icons using the Windows Program Manager.

## C4.2    OPERATION

1) The executable programs (.EXE) can be operated in the normal manner using File Manager or as icons created with Program Manager.
2) Double click on the SISO#SIM.EXE file in File Manager or its icon in the Program Manager group window to **Run** the program. A "Title Form" is displayed and a **Run** option is added to the Menu Bar.
3) Program operation is straight forward and the data input options are discussed below. Choose the "Data File" **Default** button (for default data) then the **OK** button or **Enter** key to continue operation.
4) There is a waiting period as the program generates two (2) independent random series of data points (specified by NTotal), and a form with bar chart displays progress. When completed, the first block of NData points of the input and output time series are displayed first, then *cascaded* as the autocorrelation and correlation and spectral functions are displayed.
5) **Ctrl+C** invokes a dialogue box which enables the user to terminate execution and return to the group window.

6) Use the **Run** menu options to execute a New Case or Exit from the program normally. Use the File Menu/Exit option to return to the group window.

7) The results of SISO#SIM.EXE calculations are printed to a PrintLog, which can be accessed via the Window menu option. The PrintLog results and the various plots generated by SISO#SIM.EXE can be printed directly using the File/Print menu option. The PrintLog results can also be copied to other plot programs such as MS Excel using the Windows Clipboard facility.

## C4.3  DATA INPUT AND COMPUTATIONAL PROCEDURE

The default values of the program data input are tabulated in the same format as the Data Input form as follows:

| **Time Series Data**: | | **2nd Order Process Parameters**: | |
|---|---|---|---|
| Block Data Points/Step | 100 | Static Gain ($K$) | 10 |
| Total Data Points | 4000 | Natural Frequency ($f_n$) | 1 |
| Time Step ($\Delta$ units) | 0.1 | Damping Ratio ($\xi$) | 0.7 |
| Z/X Variance Ratio | 0.2 | Time Delay | 10 |
| Method [1→3*] | 1 | | |
| Disc Store Filename | Null string (default – no disc storage) or, for example, c:\excel\sigdiags\siso.txt | | |

\* Computational Method: CF/DFT=1; TS/DFT=2; TS/FFT=3

The program default data values are invoked by pressing the DEFAULT button in the Data File form. The data input values in the bordered text boxes may be altered for each new case (Run/New Case menu option) and saved as a **filename.dat** file by pressing the SAVE AS button and typing the **filename.dat** in the Save As form that is displayed. The saved file may be selected from the list given in the Data File form by pressing the OK button.

The program first generates the input and output random sequences of 4000 data values, and then computes the process time response and associated correlation functions for the first block of 100 data values, which are displayed in cascaded graphical forms. The computations are repeated for successive 100 block data values, and the new time response and averaged correlation results are presented in the updated graphical forms, which may be selected individually and minimized or maximized or, alternatively, viewed as a *tiled* group (Window/Tile menu option). The computation terminates when the total number of data values have been processed. The correlation results are printed to the Print Log, which may be accessed via the Window/Show Print Log menu option.

# 5

# ANALYSIS OF CLOSED LOOP PROCESSES

## 5.1     INTRODUCTION AND LEARNING OBJECTIVES

Most industrial processes have inherent closed loops or feedbacks which may
be either recycle streams or, alternatively, feedback control loops whose action
is to prevent non-stationary drift in the process operating conditions, or a
combination of both. The analysis of closed loop processes poses special
problems regarding their identifiability and the accuracy of the desired process
transfer function estimates. As a consequence, closed loop processes have been
a major area of research over the past three decades or so, and this has resulted
in numerous publications on the development of several different methods of
analysis.

The purpose of this Chapter is to present an overview of closed loop
processes so that the reader is able to:

- **understand** the concepts and key criteria for the identification of closed
  loop processes from data monitored inside the loop;
- **analyse** the effects of broadband dither signals on open and closed loop
  process responses;
- **analyse** the effects of output noise/dither signal ratios on the behaviour of
  closed loop processes.

## 5.2     CLOSED LOOP PROCESSES

Consider the closed loop process depicted by the block diagram in Figure 5.1,
where $\hat{x}_t$ and $\hat{y}_t$ are the input and output measurements respectively, and $\hat{v}_t$
and $\hat{z}_t$ are random noise sources superimposed on the input and output.

The equations relating the various process variables are:

Feedforward:                $\hat{y}_t = \hat{x}_t \hat{H}(B) + \hat{z}_t$

Feedback:                   $\hat{x}_t = \hat{y}_t \hat{F}(B) + \hat{v}_t$                    ... (5.1)

where $\hat{H}(B)$ is the process weighting function ($\hat{H}(B) = \hat{h}_0 + \hat{h}_1 B + \hat{h}_2 B^2 + ...$),
$\hat{F}(B)$ is the feedback weighting function ($\hat{F}(B) = \hat{f}_0 + \hat{f}_1 B + \hat{f}_2 B^2 + ...$), and $B$

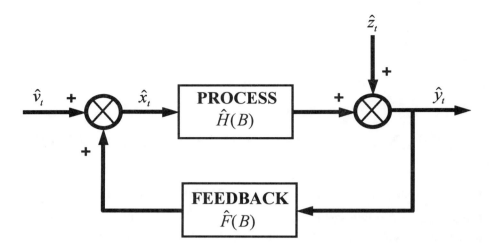

**Figure 5.1** Block diagram of a process with feedback

denotes the backward shift operator such that $B\hat{x}_t = \hat{x}_{t-1}$ (that is, $B \equiv z^{-1}$).
Eliminating $\hat{y}_t$ and $\hat{x}_t$ successively from equations (5.1) yields

$$\hat{x}_t = \left\{ \hat{z}_t \hat{F}(B) + \hat{v}_t \right\} [1 - \hat{H}(B)\hat{F}(B)]^{-1}$$

$$\hat{y}_t = \left\{ \hat{z}_t + \hat{v}_t \hat{H}(B) \right\} [1 - \hat{H}(B)\hat{F}(B)]^{-1}$$

$\qquad\qquad$ ... (5.2)

Using the usual definitions, we can compute the autocorrelation and cross-correlation functions from the equations:

$$\hat{R}_{xx}(m) = E[\hat{x}_t^* \hat{x}_{t+n}] = \left\{ \hat{R}_{zz}(m) |\hat{F}(B)|^2 + \hat{R}_{vv}(m) \right\} \left| 1 - \hat{H}(B)\hat{F}(B) \right|^{-2}$$

$$\hat{R}_{xy}(m) = E[\hat{x}_t^* \hat{y}_{t+n}] = \left\{ \hat{R}_{zz}(m) \hat{F}^*(B) + \hat{R}_{vv}(m) \hat{H}(B) \right\} \left| 1 - \hat{H}(B)\hat{F}(B) \right|^{-2} \text{ ... (5.3)}$$

$$\hat{R}_{yy}(m) = E[\hat{y}_t^* \hat{y}_{t+n}] = \left\{ \hat{R}_{zz}(m) + \hat{R}_{vv}(m) |\hat{H}(B)|^2 \right\} \left| 1 - \hat{H}(B)\hat{F}(B) \right|^{-2}$$

Equations (5.3) contain the *four* unknowns $\hat{R}_{vv}(m)$, $\hat{R}_{zz}(m)$, $\hat{H}(B)$ and $\hat{F}(B)$, and thus the process weighting function $\hat{H}(B)$ cannot be identified when only the input $\hat{x}_t$ and the output $\hat{y}_t$ signals are measured. It follows from the first equation in (5.1) that an alternative form of the input-output cross-correlation function $\hat{R}_{xy}(m)$ is given by:

$$\hat{R}_{xy}(m) = \hat{R}_{xx}(m)\hat{H}(B) + \hat{R}_{xz}(m)$$

$\qquad\qquad$ ... (5.4)

and substitution of equations (5.3) in (5.4) yields, with some rearrangement, the solutions

$$\hat{H}(B)\hat{F}(B) = 1 \qquad \qquad \text{... (5.5}a\text{)}$$

and

$$\hat{R}_{zz}(m)\hat{F}^*(B) = \hat{R}_{xz}(m)\left\{1 - \hat{H}^*(B)\hat{F}^*(B)\right\} \qquad \text{... (5.5}b\text{)}$$

Equation (5.5a) states that, for all $B$, the process function $\hat{H}(B)$ is the inverse of the feedback function $\hat{F}(B)$, and is termed a *degenerate process*. Equation (5.5b), or its equivalent complex conjugate form:

$$\hat{R}_{zx}(m) = \hat{R}_{zz}(m)\hat{F}(B)\left\{1 - \hat{H}(B)\hat{F}(B)\right\}^{-1} \qquad \text{... (5.5}c\text{)}$$

gives a measure of the effect of the output residual noise on the input measurement $\hat{x}(t)$. In the *open loop* case, $\hat{F}(B)$ is zero, and thus $\hat{R}_{zx}(m) = 0$, so that equation (5.4) can be used for identifying the open loop transfer function in the normal manner. Clearly, this constraint no longer applies for closed loop processes, since the input and the output residual noise are now correlated via the feedback loop and the direct application of the correlation/spectral analysis methods discussed in the previous Chapters yield erroneous estimates of the process weighting and transfer functions.

Before considering the criteria by which closed loop processes are identified, we will consider the more general case encountered in practice in which the input and output noise terms, $\hat{v}_t$ and $\hat{z}_t$ respectively, are 'coloured noise' (filtered white noise), as depicted by the block diagram in Figure 5.2. The input and output noise terms are now respectively

$$\hat{v}_t = \hat{C}(B)\hat{\eta}_t; \quad \hat{z}_t = \hat{D}(B)\hat{\epsilon}_t \qquad \text{... (5.6)}$$

where $\hat{C}(B)$ and $\hat{D}(B)$ may be known or, as is more often the case in practice, unknown filtering operations on the input and output noise sources, respectively. Hence, equations (5.3) become:

$$\hat{R}_{xx}(m) = \text{E}[\hat{x}_t^* \hat{x}_{t+n}] = \left\{\hat{\sigma}_\eta^2 |\hat{D}(B)|^2 |\hat{F}(B)|^2 + \hat{\sigma}_\epsilon^2 |\hat{C}(B)|^2\right\}\left|1 - \hat{H}(B)\,\hat{F}(B)\right|^{-2}$$

$$\hat{R}_{xy}(m) = \text{E}[\hat{x}_t^* \hat{y}_{t+n}] = \left\{\hat{\sigma}_\eta^2 |\hat{D}(B)|^2 \,\hat{F}^*(B) + \hat{\sigma}_\epsilon^2 |\hat{C}(B)|^2 \,\hat{H}(B)\right\}\left|1 - \hat{H}(B)\,\hat{F}(B)\right|^{-2}$$

$$\hat{R}_{yy}(m) = \text{E}[\hat{y}_t^* \hat{y}_{t+n}] = \left\{\hat{\sigma}_\eta^2 |\hat{D}(B)|^2 + \hat{\sigma}_\epsilon^2 |\hat{C}(B)|^2 |\hat{H}(B)|^2\right\}\left|1 - \hat{H}(B)\,\hat{F}(B)\right|^{-2}$$

$$\text{... (5.7)}$$

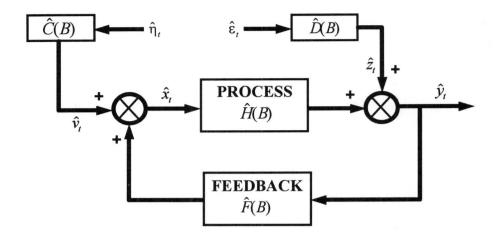

**Figure 5.2** Block diagram of a closed loop process with coloured noise

where $\hat{\sigma}_\eta^2 = E[(\hat{\eta}_t - \hat{\bar{\eta}})^2]$ and $\hat{\sigma}_\varepsilon^2 = E[(\hat{\varepsilon}_t - \hat{\bar{\varepsilon}})^2]$.

Following the (frequency domain) analysis by Priestley (1981), we now define the function $\hat{\Gamma}(B)$ from equation (5.4) such that

$$\hat{R}_{xz}(m) = \hat{R}_{xx}(m)\hat{\Gamma}(B) = \hat{R}_{xy}(m) - \hat{R}_{xx}(m)\hat{H}(B) \qquad \text{... (5.8)}$$

and on substitution for $\hat{R}_{xx}(m)$ and $\hat{R}_{xy}(m)$ from equations (5.7) we obtain

$$\hat{\Gamma}(B) = \frac{\hat{\sigma}_\varepsilon^2 |\hat{D}(B)|^2 \hat{F}^*(B)\{1 - \hat{H}(B)\hat{F}(B)\}}{\hat{\sigma}_\varepsilon^2 |\hat{D}(B)|^2 |\hat{F}(B)|^2 + \hat{\sigma}_\eta^2 |\hat{C}(B)|^2} \qquad \text{... (5.9)}$$

We will use this equation to investigate the criteria for identifying closed loop processes.

**5.3      IDENTIFIABILITY CRITERIA**

I)   **Degenerate Process $\hat{F}(B) = \hat{H}^{-1}(B)$**

In this case $\hat{\Gamma}(B) \equiv 0$ and, as we have seen, this results in a *degenerate process*. This degeneracy arises from the fact that, from equations (5.2), $\hat{F}(B) = \hat{H}^{-1}(B) \Rightarrow \hat{\sigma}_\varepsilon^2 \hat{D}(B) = -\hat{\sigma}_\eta^2 \hat{C}(B)\hat{H}(B)$, which contradicts the assumption that $\hat{\eta}_t$ and $\hat{\varepsilon}_t$ are uncorrelated white noise processes. Note, however, that since $\hat{\Gamma}(B)$ is a measure of the bias in the process transfer function estimates, equation (5.8), the term $[1 - \hat{H}(B)\hat{F}(B)]$ should ideally be *small* for all $B$ (frequencies).

## II) No Dither Signal $\hat{C}(B) \equiv 0$

This corresponds to a process where the feedback loop is *noise-free*, and equation (5.9) becomes:

$$\hat{\Gamma}(B) = \hat{F}^{-1}(B) - \hat{H}(B) \qquad \dots (5.10)$$

and substituting this expression into equation (5.8) yields

$$\hat{R}_{xy}(m) = \hat{R}_{xx}(m)\hat{F}^{-1}(B) \qquad \dots (5.11)$$

that is, the process weighting function estimates are equivalent to the inverse of the feedback transfer function, and gives no information about the process weighting function $\hat{H}(B)$. Furthermore, since $\hat{H}(B) = \hat{h}_0 + \hat{h}_1 B + \dots + \hat{h}_K B^K$ and it is reasonable to assume for most industrial processes that $\hat{h}_0 = 0$ (no instantaneous transfer of the input to the output), then it follows from equations (5.11) that *a substantial sample cross-correlation at zero lag ($m = 0$) usually indicates the presence of a feedback loop*.

## III) No Feedback $\hat{F}(B) \equiv 0$

This is equivalent to the open loop case ($\hat{\Gamma}(B) \equiv 0$), and we see from equation (5.9) that, as expected,

$$\hat{R}_{xy}(m) = \hat{R}_{xx}(m)\hat{H}(B) \qquad \dots (5.12)$$

However, this condition ($\hat{\Gamma}(B) \equiv 0$) also corresponds to the degenerate case I above, and an additional requirement is that $\hat{\sigma}_\eta^2 |\hat{C}(B)|^2 \neq 0$ for all $B$ (frequencies).

## IV) The Output/Input Noise Ratio ($\hat{\sigma}_\varepsilon^2 / \hat{\sigma}_\eta^2$) is Small

It is clear from equation (5.9) that $\hat{\Gamma}(B)$ is small as $\hat{\sigma}_\varepsilon^2 / \hat{\sigma}_\eta^2 \to 0$, and thus the resulting process weighting function estimates are relatively unbiased. This assumption is reasonable when a *dither signal* is superimposed on the process or, in certain cases, when a process is under manual control; that is, when the 'noise' in the feedback loop is much larger than the 'noise' inherent in the process.

The application of a broadband dither signal $(\hat{\eta}_t)$ enables the process transfer function to be identified from the feedforward equation in (5.1):

$$\hat{R}_{\eta y}(k) = \hat{R}_{\eta x}(k)\hat{H}(B) \qquad 0 \le k \le M \qquad \qquad \dots (5.13)$$

where it has been assumed that $\hat{R}_{\eta \varepsilon}(k) = 0, k \ge 0$. Similarly, the feedback equation in (5.1) yields the feedback estimates:

$$\hat{R}_{\eta x}(k) = \hat{R}_{\eta y}(k)\hat{F}(B) + \hat{R}_{\eta \eta}(k)C(B) \qquad 0 \le k \le M \qquad \dots (5.14)$$

provided that the input filter function $C(B)$ is known. As we have seen, we ideally require the dither signal to have the following properties:

$$\begin{aligned} \hat{R}_{\eta \eta}(k) &= \hat{\sigma}_{\eta}^2 & k &= 0 \\ \hat{R}_{\eta \eta}(k) &= 0 & k &> 0 \end{aligned} \qquad \dots (5.15)$$

and hence $C(B) = c_0 = 1$. This broadband property of the dither signal is often termed 'frequency richness', and is designed, where practical, to enable *all* frequency modes of the process weighting function $\hat{H}(B)$ to be identified. A variety of computer generated dither signals are employed in the adaptive control applications to obtain reliable estimates of the process (hence control) parameters or, in some applications, to keep the adaptive algorithm 'active'; that is, to prevent it from going to 'sleep' due to a lack of dynamic information ('persistent excitation'). The most common dither signal normally used in practice is the pseudo random binary sequence (prbs), but ternary and higher order sequences are also used for their sharper cutoff characteristics near the Nyquist frequency.

We will examine this case in more detail in the simulation example in the following section.

## 5.4      A CLOSED LOOP SIMULATION EXAMPLE

We will now illustrate various aspects of the above criteria using the closed loop simulation program (LOOP#SIM.EXE) supplied with this text. The program and its installation and operation is described in Appendix C5. In the following discussion, we will confine our remarks to salient features of closed loop systems.

### 5.4.1      Open Loop Test Process Characteristics

The test process is the discrete equivalent of a continuous second order process with the following key parameters:

**Analysis Parameters**:
Data Points/Block  = 100
Total Data Points  = 2000
Number of Lags  = 50
Time Interval  = 0.1 min
Dither Signal Gain  = 1
Noise Signal Gain  = 0→32

**Process Parameters**:
Static Gain ($K$)  = 10
Natural Frequency ($f_n$)  = 1 cpm
Damping Ratio ($\xi$)  = 0.70
Time Delay  = 0

Calculation of the equivalent discrete process parameters using the methods discussed in Chapter 3 is given in Appendix B5. The open loop correlations functions are computed using SISOLOOP.EXE and are presented in Figure 5.3. The process is excited by white noise low-pass filtered below 4 cpm, and this gives an input autocorrelation with a characteristic spike at the zero lag axis (Figure 5.3$a$). The narrow band output autocorrelation and input-output cross-correlation are given in Figures 5.3$b$ and 5.3$c$ respectively. The off-set of the maximum from the zero lag axis in the latter case is due entirely to process 'inertia' in this example, but in more general process analysis it may include a further off-set due to an inherent time delay. It can be shown (see Tutorial Exercise 5.1) that the maximum occurs at time $t_m$ where, to the nearest time step,

$$t_m = \frac{\tan^{-1}\sqrt{1-\xi^2}\big/\xi}{\omega_n\sqrt{1-\xi^2}} = 0.177\,\text{min} \approx 0.2\,\text{min} \qquad ...\,(5.16)$$

in which $\omega_n$ is the natural frequency in radians/minute and $\xi$ is the damping ratio. (We note in passing that the discrete process is slightly 'non-minimum phase', and that modified Z-transform methods could be used to obtain more accurate results.)

### 5.4.2 Closed Loop Process Characteristics

We now investigate what happens to the input autocorrelation and cross-correlation functions when the process is placed in closed loop with a minimum variance controller, the parameters of which are evaluated in Appendix B5. Figure 5.4 presents simulation results in which broadband dither and output noise signals are superimposed on the closed loop process for selected noise/dither signal variance ratios ranging from zero to 1000; that is, $\hat{\sigma}_\varepsilon^2/\hat{\sigma}_\eta^2 = 0, 1, 10, 100, 1000$. We note the following observations:

1) *Input Autocorrelations*: The input autocorrelation for zero variance ratio ($\sigma_\varepsilon^2/\sigma_\eta^2 = 0$, Figure 5.4$a$) is similar to the open loop case except that it now includes some non-zero information at the base of the 'spike' at the

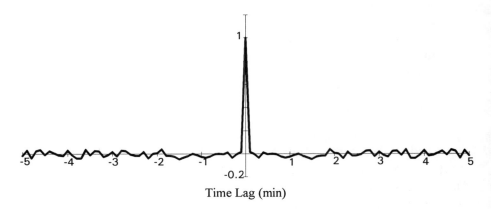

Time Lag (min)

*a*) Input autocorrelation $R_{xx}(\tau)$

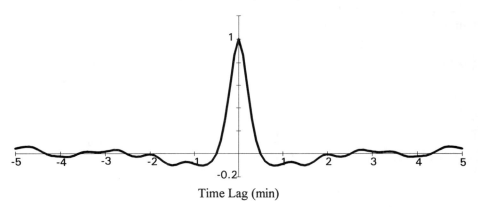

Time Lag (min)

*b*) Output autocorrelation $R_{yy}(\tau)$

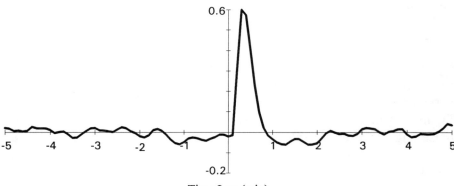

Time Lag (min)

*c*) Input-output cross-correlation $R_{xy}(\tau)$

**Figure 5.3** Process correlation functions (no feedback)

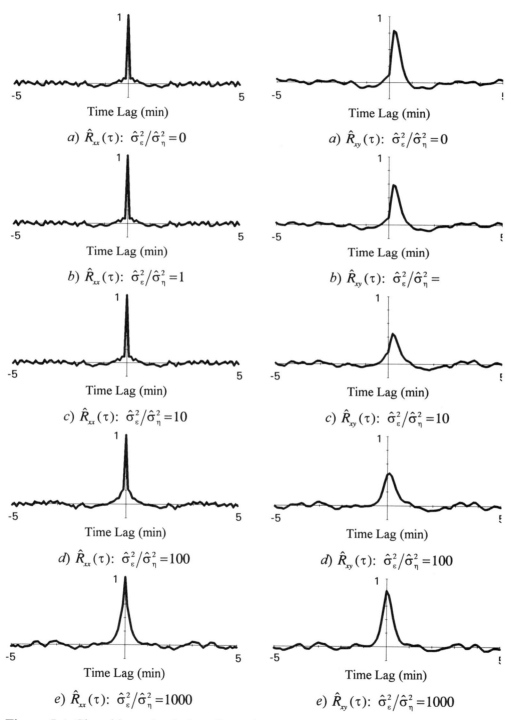

a) $\hat{R}_{xx}(\tau)$: $\hat{\sigma}_{\varepsilon}^2/\hat{\sigma}_{\eta}^2 = 0$

a) $\hat{R}_{xy}(\tau)$: $\hat{\sigma}_{\varepsilon}^2/\hat{\sigma}_{\eta}^2 = 0$

b) $\hat{R}_{xx}(\tau)$: $\hat{\sigma}_{\varepsilon}^2/\hat{\sigma}_{\eta}^2 = 1$

b) $\hat{R}_{xy}(\tau)$: $\hat{\sigma}_{\varepsilon}^2/\hat{\sigma}_{\eta}^2 =$

c) $\hat{R}_{xx}(\tau)$: $\hat{\sigma}_{\varepsilon}^2/\hat{\sigma}_{\eta}^2 = 10$

c) $\hat{R}_{xy}(\tau)$: $\hat{\sigma}_{\varepsilon}^2/\hat{\sigma}_{\eta}^2 = 10$

d) $\hat{R}_{xx}(\tau)$: $\hat{\sigma}_{\varepsilon}^2/\hat{\sigma}_{\eta}^2 = 100$

d) $\hat{R}_{xy}(\tau)$: $\hat{\sigma}_{\varepsilon}^2/\hat{\sigma}_{\eta}^2 = 100$

e) $\hat{R}_{xx}(\tau)$: $\hat{\sigma}_{\varepsilon}^2/\hat{\sigma}_{\eta}^2 = 1000$

e) $\hat{R}_{xy}(\tau)$: $\hat{\sigma}_{\varepsilon}^2/\hat{\sigma}_{\eta}^2 = 1000$

**Figure 5.4** Closed loop simulations for various output noise/dither variance ratios

zero time lag axis due to feedback effects. The amplitude of this feedback information remains relatively unchanged until the variance ratios increase to 100 and 1000 (Figures 5.4$d$ and 5.4$e$) respectively. In the latter case, the shape of the input autocorrelation is very similar to the open loop output autocorrelation (Figure 5.3$b$), which confirms that the (filtered) output signal is now driving the closed loop process.

2)  *Cross-correlations*: As expected, the feedback effects have a profound influence on the cross-correlation results. Even at zero variance ratio, there is a moderately high cross-correlation amplitude at the zero time lag axis, and this amplitude increases as the peak in the cross-correlation function moves, as expected, towards the left half plane at the higher variance ratios (Figures 5.4$d$ and 5.4$e$). The end result is that, consistent with the theoretical concepts discussed above, the cross-correlation function of the closed loop process no longer satisfies the *physical realisability* condition and, as a consequence, the process transfer (or weighting) function cannot be identified solely from measurements of the input $x(t)$ and output $y(t)$.

3)  *Process Identification*: In practical terms this means that processes which have large recycle streams (e.g. flotation circuits in mineral processing plants) or inherent feedback mechanisms (e.g. in biology and medicine) are difficult to identify in practice, and the bias errors in the process transfer function estimates may be sufficiently large to yield unreliable causal interactions for diagnostic applications.

4)  *Process Control*: Likewise, in control applications (particularly adaptive control applications), it is important that, as noted above, the dither signal has sufficient broadband energy to 'persistently excite' all frequency modes of the process dynamics. Otherwise, the bias errors in the process parameter estimates and the associated (adaptive) control parameter estimates may be sufficiently large to drive the closed loop process unstable. (Further discussion on control applications is beyond the scope of this book, and interested readers are referred to the wide range of excellent control texts for further information.)

## REFERENCES

Astrom, K. J., & Wittenmark, B., 1984, *Computer-Controlled Systems: Theory and Design*, Prentice-Hall, Englewood Cliffs.

Isermann, R., 1985, *Digital Control Systems*, Springer-Verlag, Berlin.

Priestley, M. B., 1981, *Spectral Analysis and Time Series*, **1** & **2**, Academic Press, London.

## TUTORIAL PROBLEMS

1) For a second order process given by the transfer function equation:

$$H(s) = \frac{Y(s)}{X(s)} = \frac{Ke^{-s\tau_d}}{s^2 + 2\xi\omega_n s + \omega_n^2}$$

prove that the time taken for the impulse response of the process to reach its maximum value is given by equation (5.16); that is,

$$t_m = \frac{\tan^{-1}\sqrt{1 - \xi^2}/\xi}{\omega_n\sqrt{1 - \xi^2}}$$

2) Using equations (B5.9) and (B5.11), prove that the minimum variance controller parameters are given by:

a) No Delay ($d = 0$)

   Autoregressive:    $p_0 = r/b_1 \quad p_1 = b_1 \quad p_2 = b_2$

   Moving Average:    $q_0 = -a_1 \quad q_1 = -a_2$

b) With Delay ($d = 1$)

   Autoregressive:    $p_0 = r/b_1 \quad p_1 = b_1 \quad p_2 = b_2 - a_1 b_1 \quad p_3 = -a_1 b_2$

   Moving Average: $q_0 = a_1^2 - a_2 \quad q_1 = a_1 a_2$

and so obtain the values given in equations (B5.12) and (B5.13). What observations can you make about the action of the minimum variance controller on the process?

3) Using the LOOP#SIM.EXE simulation program, investigate one or more of the following closed loop processes:

a) the test process over the same range of noise/dither signal variance ratios ($\sigma_\varepsilon^2/\sigma_\eta^2$), but with the data points/step = 50 and total data points = 500;

b) repeat Exercise 3a for processes with damping ratios ($\xi$) ranging from 0→0.99 and zero time delay ($d = 0$);

c) repeat Exercise 3b for unit time delay ($d = 1$).

In each case, list your observations about the resulting closed loop process response. How do these compare with the corresponding open loop process response (Loop = 0)?

NOTE: The results of LOOP#SIM.EXE calculations are printed to a Print Log, which can be accessed via the Window/Open Print Log menu option. The Print Log results and the various plots generated by the program can be printed directly using the File/Print menu option. The Print Log results can also be copied to other plot programs such as MS Excel using the Windows Clipboard facility.

### SOLUTIONS TO SELECTED TUTORIAL PROBLEMS

1)  Taking the inverse Laplace transforms of $H(s)$, we obtain the time response to an unit impulse as:

$$h(t) = \frac{K}{\omega_n\sqrt{1-\xi^2}} e^{-\xi\omega_n t} \sin\left(\omega_n\sqrt{1-\xi^2}\, t\right)$$

Differentiating this equation with respect to $t$ we obtain:

$$\frac{\partial h(t)}{\partial t} = \frac{K}{\sqrt{1-\xi^2}} e^{-\xi\omega_n t}\left[\sqrt{1-\xi^2}\,\cos\left(\omega_n\sqrt{1-\xi^2}\right) - \xi \sin\left(\omega_n\sqrt{1-\xi^2}\right)\right]$$

at the maximum. The term in the square brackets yields the desired result. (Note the alternative asymptotic solution $e^{-\xi\omega_n t} = 0 \rightarrow t = \infty$.)

2)  *a)*    No delay ($d = 0$):

For a second order zero delay discrete process $A(z) = 1 + a_1 z^{-1} + a_2 z^{-2}$, $B(z) = b_1 z^{-1} + b_2 z^{-2}$ and $D(z) = F(z) = 1$, equation (B5.11). Substituting these values in equation (B5.3) yields $z^{-1}L(z) = 1 - A(z)$, and equating coefficients in the powers of $z$ yields $l_0 = -a_1 ; l_1 = -a_2$. Equation (B5.9) gives $Q(z) = L(z)$, hence $q_0 = -a_1 ; q_1 = -a_2$, and $P(z) = B(z) + r/b_1$, hence $p_0 = r/b_1 ; p_1 = b_1 ; p_2 = b_2$.

*b)*    With delay ($d = 1$):

In this case $B(z) = b_1 z^{-2} + b_2 z^{-3}$, $D(z) = 1 + d_1 z^{-1}$ and $F(z) = 1 + f_1 z^{-1}$, and the desired results are obtained using the same procedure as in case (*a*).

# APPENDIX A5

## PROCESS PARAMETERS AND FILTER COEFFICIENTS

### A5.1 Process Parameters

The open loop process is assumed to be a second order continuous process with a transfer function of the form (cf. equations (4.40) to (4.44) in section 4.7.3, Chapter 4):

$$H(s) = \frac{Y(s)}{X(s)} = \frac{Ke^{-s\tau_d}}{s^2 + 2\xi\omega_n s + \omega_n^2} \qquad \ldots (A5.1)$$

where

    the static gain          $K = 10$

    the damping ratio     $\xi = 0.70$

    the natural frequency  $f_n = 1$ cycles/min$\rightarrow\omega_n = 2\pi f_n = 6.28319$ rad/min

    the process time delay $\tau_d = 0$

The process time constants and associated coefficients are:

$$T_1 = \frac{e^{j\lambda}}{\omega_n} = 0.1114 + j0.1137 \rightarrow S_1 = e^{-\Delta/T_1} = 0.5804 + j0.2794$$

$$\ldots (A5.2)$$

$$T_2 = \frac{e^{-j\lambda}}{\omega_n} = 0.1114 - j0.1137 \rightarrow S_2 = e^{-\Delta/T_2} = 0.5804 - j0.2794$$

where $\lambda = \cos(\xi)$. The equivalent discrete process transfer function is given by (see Appendix B3, Chapter 3):

$$H(z) = \frac{B(z)}{A(z)} = \frac{(b_1 z^{-1} + b_2 z^{-2})}{(1 + a_1 z^{-1} + a_2 z^{-2})} \qquad \ldots (A5.3)$$

which is equivalent to the discrete [ARMA(2,2)] process difference equation:

$$y_t = -a_1 y_{t-1} - a_2 y_{t-2} + b_1 x_{t-1} + b_2 x_{t-2}; \quad t \geq 1$$
$$= 0; \qquad\qquad\qquad\qquad\qquad\qquad t \leq 0 \qquad \ldots (A5.4)$$

The discrete parameters and their calculated values are given by:

$$a_1 = -(S_1 + S_2) = -1.1608$$
$$a_2 = S_1 S_2 \qquad = 0.4149 \qquad \ldots (A5.5a)$$

$$b_1 = \frac{K}{T_1 - T_2}[T_1(1-S_1) - T_2(1-S_2)] \qquad = 1.4572$$

$$\qquad \qquad \qquad \qquad \qquad \qquad \qquad \qquad \qquad \qquad \qquad \text{... (A5.5$b$)}$$

$$b_2 = \frac{-K}{T_1 - T_2}[T_1(1-S_1)S_2 - T_2(1-S_2)S_1] = 1.0845$$

## A5.2    Lowpass Digital Filter Coefficients

The random dither and output noise signals are filtered using a lowpass digital filter of the form:

Dither signal:  $v_t = Rv_{t-1} + G(\eta_t + \eta_{t-1})$

Noise signal:  $z_t = Rz_{t-1} + G(\varepsilon_t + \varepsilon_{t-1})$ $\qquad$ ... (A5.6)

where

$G$ = static gain $\qquad \qquad = (1-R)/2$

$R$ = cutoff frequency ratio $\quad = 1 - 2f_L/f_N$ $\qquad$ ... (A5.7)

Thus

$$C(z) = \frac{G(1+z^{-1})}{1 - Rz^{-1}} \qquad \qquad \text{... (A5.8)}$$

and since the input and output digital filters are assumed to be identical, $D(z) = C(z)$.

The dither and noise signals were filtered with a lowpass cutoff frequency of $f_L = 0.80 f_N$, so that $R = -0.60$ and $G = 0.80$. The frequency response modulus, $|C(f)|$, for these values is given by the second upper curve in Figure 5.5.

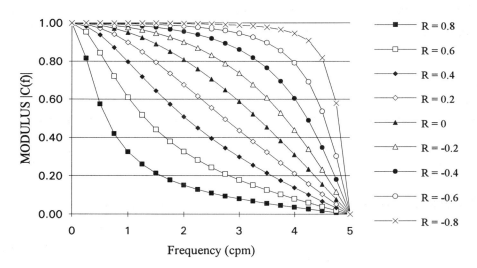

**Figure 5.5** Lowpass digital filter frequency response for various $r$ values

# APPENDIX B5

## MINIMUM VARIANCE CONTROL

### B5.1    Process Model

The process is modelled by a parametric autoregressive moving average ARMA(m,m,d) model of the form:

$$A(z)y_t = B(z)z^{-d}x_t + D(z)\varepsilon_t \qquad \ldots \text{(B5.1)}$$

### B5.2    *d*-Step Ahead Predictor

$$z^{d+1}y_t = \frac{B(z)}{A(z)}zx_t + \frac{D(z)}{A(z)}z^{d+1}\varepsilon_t$$

$$= \frac{B(z)}{A(z)}zx_t + \left\{F(z)z^{d+1} + \frac{L(z)}{A(z)}\right\}\varepsilon_t \qquad \ldots \text{(B5.2)}$$

$$\text{or} \quad y_{t+d+1} = \frac{B(z)}{A(z)}zx_t + \frac{L(z)}{A(z)}\varepsilon_t + F(z)\varepsilon_{t+d+1}$$

where

$$D(z) = F(z)A(z) + z^{-(d+1)}L(z) \qquad \ldots \text{(B5.3)}$$

$F(z)\varepsilon_{t+d+1}$ is the noise component that **cannot** be controlled, and $(L(z)\varepsilon_t)/A(z)$ is the component that **can** be controlled. Now, from (B5.1)

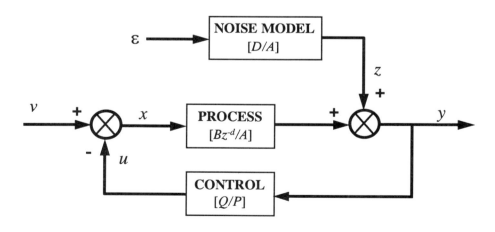

**Figure 5.7** Block diagram of closed loop process with minimum variance controller

$$\varepsilon_t = \frac{A(z)}{D(z)} y_t - \frac{B(z)}{D(z)} z^{-d} x_t \qquad \text{... (B5.4)}$$

Substitute (B5.4) into (B5.2):

$$y_{t+d+1} = \frac{B(z)}{A(z)} z x_t + \frac{L(z)}{A(z)} \left[ \frac{A(z)}{D(z)} y_t - \frac{B(z)}{D(z)} z^{-d} x_t \right] + F(z)\varepsilon_{t+d+1}$$

$$= \frac{L(z)}{D(z)} y_t + \frac{B(z)F(z)}{D(z)} x_t + F(z)\varepsilon_{t+d+1} \qquad \text{... (B5.5)}$$

## B5.3     Control Function

$$J_{t+d+1} = E\left[ y_{t+d+1}^2 + r u_t^2 \right]$$

$$= E\left[ (F(z)\varepsilon_{t+d+1})^2 \right] + E\left[ \left\{ \frac{L(z)}{D(z)} y_t + \frac{B(z)F(z)}{D(z)} x_t \right\}^2 + r(v_t - x_t)^2 \right]$$

$$\qquad \text{... (B5.6)}$$

where $r$ is a weighting factor which limits the magnitude of the controller signal $u_t$ and the first and second terms in equation (B5.6) are uncorrelated.

Differentiating (B5.6) with respect to $x_t$ gives:

$$J'_{t+d+1} = \frac{\partial J_{t+d+1}}{\partial x_t} = E\left[ b_1 \left\{ \frac{L(z)}{D(z)} y_t + \frac{B(z)F(z)}{D(z)} x_t \right\} - r(v_t - x_t) \right] \qquad \text{... (B5.7)}$$

$$= 0$$

Equation (B5.7) is a minimum when

$$\left\{ L(z) y_t + B(z)F(z) x_t \right\} - \frac{rD(z)}{b_1}(v_t - x_t) = 0$$

that is,

$$x_t = \frac{\dfrac{rD(z)}{b_1} v_t - L(z) y_t}{\left[ B(z)F(z) + \dfrac{rD(z)}{b_1} \right]} \qquad \text{... (B5.8)}$$

Let

$$Q(z) = L(z) = z^{d+1}\{D(z) - F(z)A(z)\}$$
$$P(z) = \{B(z)F(z) + rD(z)/b_1\} \qquad \text{(B5.9)}$$

then

$$x_t = \frac{\dfrac{rD(z)}{b_1}v_t - Q(z)y_t}{P(z)} \qquad \ldots \text{(B.10)}$$

where

$$
\begin{aligned}
A(z) &= 1 + a_1 z^{-1} + \ldots + a_m z^{-m} \\
B(z) &= \quad b_1 z^{-1} + \ldots + b_m z^{-m} \\
D(z) &= 1 + d_1 z^{-1} + \ldots + d_m z^{-m} \\
F(z) &= 1 + f_1 z^{-1} + \ldots + f_d z^{-d} \qquad 0 \le d < m \qquad \ldots \text{(B5.11)} \\
L(z) &= 1 + l_1 z^{-1} + \ldots + l_{m-1} z^{m-1} \\
P(z) &= p_0 + p_1 z^{-1} + \ldots + p_{m+d} z^{-m-d} \\
Q(z) &= q_0 + q_1 z^{-1} + \ldots + q_{m-1} z^{-m+1}
\end{aligned}
$$

The coefficients of the functions $P(z)$ and $Q(z)$ in equations (B5.9) are derived by substituting the polynomial functions from equations (B5.11) and equating the coefficients for the same powers in $z$ (see Exercise 5.2).

Note that minimum variance controllers can only give better control performance with an uncontrolled pure deadtime process if the disturbance signal $\varepsilon_t$ acting on $y_t$ is an ARMA or MA process of order $m \ge d$, or $d < m$.

## B5.4 Minimum Variance Controller Parameters

Using the model parameter values given in equation (A5.5), the minimum variance control parameters computed by SISOLOOP.EXE (see Print Log) are, for $m = 2$, $D(z) = 1$:

i) *No Delay* $(d = 0)$: $F(z) = f_0 = 1 \rightarrow f_0 = 1$
  Autoregressive:
  $$p_0 = r/b_1 = 34.3124; \; p_1 = b_1 = 1.4572; \; p_2 = b_2 = 1.0845 \qquad \ldots \text{(B5.12a)}$$
  Moving average:
  $$q_0 = -a_1 = 1.1608; \; q_1 = -a_2 = -0.4149 \qquad \ldots \text{(B5.12b)}$$

ii) *With Delay* $(d = 1)$: $F(z) = f_0 + f_1 z^{-1} = 1 - a_1 z^{-1} \rightarrow f_0 = 1; \; f_1 = -a_1$
  Autoregressive:
  $$p_0 = r/b_1 = 34.3124; \; p_1 = b_1 = 1.4572; \; p_2 = b_2 - a_1 b_1 = 2.7760$$
  $$p_3 = -a_1 b_2 = 1.2589 \qquad \ldots \text{(B5.13a)}$$
  Moving average:
  $$q_0 = a_1^2 - a_2 = 0.9325; \; q_1 = a_1 a_2 = -0.4816 \qquad \ldots \text{(B5.13b)}$$

# APPENDIX C5

## LOOP#SIM PROGRAM INSTALLATION AND OPERATION

LOOP#SIM.EXE computes the open loop and closed loop correlation functions for a second order process for a range of noise/input signal ratios. It has been designed to run under Windows 3.1 with video display settings of 640×480 pixels×256 colours, and also runs under Windows 95 with the same video display settings. Refer to the **readme.txt** document for further information.

### C5.1     INSTALLATION

1) Insert the floppy disc in the FDD and type **a:\setup filename** in the Windows 3.1 Program Manager File/RUN menu.
2) The **setup.bat** file copies *all* the files on the floppy disc to the hard drive directory **c:\filename**. (NOTE: The setup procedure will terminate if **filename** is not specified.)
3) The **setup.bat** file also decompresses the RUN library files **cable.dll** and **rlzrun20.rts** and instals them into a newly created Windows sub-directory **c:\windows\rlzrun20**.
4) For ease of operation, it is recommended that the four simulation programs in the **c:\filename** directory be installed as a program group icon with their separate program icons using the Windows Program Manager.

### C5.2     OPERATION

1) The executable programs (.EXE) can be operated in the normal manner using File Manager or as icons created with Program Manager.
2) Double click on the LOOP#SIM.EXE file in File Manager or its icon in the Program Manager group window to **Run** the program. A Title Form is displayed and a **Run** option is added to the Menu Bar.
3) Program operation is straight forward and the Data File input options are discussed below. Choose either the **Default** button (for default data) or select a file from the list and then press the **OK** button or **Enter** key to continue operation.
4) There is a waiting period as the program generates two (2) independent random series of data points (specified by NTotal), and a form with bar chart displays progress. When completed, the first block of NData points of the input and output time series are displayed first, then *cascaded* as the autocorrelation and correlation and spectral functions are displayed.
5) **Ctrl+C** invokes a dialogue box which enables the user to terminate execution and return to the group window.

6) Use the **Run** menu options to execute a New Case or Exit from the program normally. Use the File Menu/Exit option to return to the group window.

7) The results of LOOP#SIM.EXE calculations are printed to a PrintLog, which can be accessed via the Window menu option. The PrintLog results and the various plots generated by LOOP#SIM.EXE can be printed directly using the File/Print menu option. The PrintLog results can also be copied to other plot programs such as MS Excel using the Windows Clipboard facility.

## C5.3    DATA INPUT AND COMPUTATIONAL PROCEDURE

The default values of the program data input are tabulated in the same format as the Data Input form as follows:

| | | PROCESS PARAMETERS: | |
|---|---|---|---|
| Block Data Points/Step | 100 | | |
| Total Data Points | 1000 | Static Gain ($K$) | 10 |
| Time Step ($\Delta$ units) | 0.1 | Natural Frequency ($f_n$) | 1 |
| Number of Lags ($M$) | 51 | Damping Ratio ($\xi$) | 0.7 |
| Loop Option [OL=0; CL=1] | 0 | Time Delay [0 or 1] | 1 |
| Dither Signal Gain | 1 | | |
| Noise Signal Gain | 0 | | |
| Disc Store Filename | | Null string (default – no disc storage) or, for example, c:\excel\sigdiags\loop.txt | |

The program default data values are invoked by pressing the **Default** button in the Data File form. The data input values in the bordered text boxes may be altered for each new case (Run/New Case menu option) and saved as a **filename.dat** file by pressing the SAVE AS button and typing the **filename.dat** in the Save As form that is displayed. The saved file may be selected from the list given in the Data File form by pressing the OK button.

The program first generates the input and output random sequences of 1000 data values, and then computes the process time response and associated correlation functions for the first block of 100 data values, which are displayed in cascaded graphical forms. The computations are repeated for successive 100 block data values, and the new time response and averaged correlation results are presented in the updated graphical forms, which may be selected individually and minimized or maximized or, alternatively, viewed as a *tiled* group (Window/Tile menu option). The computation terminates when the total number of data values have been processed. The correlation results are printed to the Print Log, which may be accessed via the Window/Show Print Log menu option.

# 6

## PARAMETRIC SPECTRAL ANALYSIS

### 6.1    INTRODUCTION AND LEARNING OBJECTIVES

The Fourier transform techniques that we have used so far require long data records for reliability of the spectral estimates, both in terms of its bias and frequency resolution. This has led to the development of alternative *parametric methods* to investigate the frequency characteristics of short data records, such as those encountered in the areas of materials science, geophysics and nuclear engineering, where *burst phenomena* (i.e. signals with their energy concentrated in short bursts of time) are often encountered. Such signals have been found to be more amenable to analysis by parametric (usually autoregressive) prediction error filters, from which the power spectra and cross-spectra may be determined. We review these techniques here because of their growing importance in the field of digital signal processing.

In this Chapter, we:

- review the types of discrete parameter models used for power spectral density estimation;
- establish the relationship between the observed *autocorrelation* sequence, $R_{xx}(k)$, and autoregressive moving-average (ARMA) model parameters;
- describe the Yule-Walker method for estimating the AR parameters;
- determine the power spectral density estimates of short data records using maximum entropy spectral analysis (MESA) methods;
- determine the 'optimum' model order using various criteria;
- perform simulation experiments using the computer program supplied.

The following analyses assumes that readers are familiar with the principles of sampled data processes discussed in Chapter 3, particularly section 3.2.4.

### 6.2    PARAMETRIC MODELS

#### 6.2.1    Problem Statement

Suppose that we have a weakly stationary random processs characterised by the sequence $\{x_t\}$, as shown in Figure 6.1. Our objective is to determine the

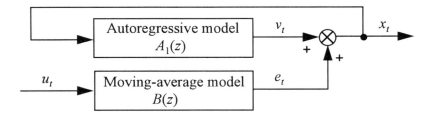

**Figure 6.1** Simulation of a 'true' signal

'optimum' parameters of a causal model which can track the current value, $x_t$, with minimum error $e_t$. Referring to Figure 6.1, we see that the current value $x_t$ is given by:

$$x_t = v_t + e_t \qquad \qquad \text{... (6.1)}$$

where $v_t$ is the *deterministic* component which is recursively related to the output signal by the autoregressive (AR) model:

$$v_t = -\sum_{j=1}^{p} \alpha_j x_{t-j} = -A_1(z)x_t . \qquad \qquad \text{... (6.2)}$$

and $e_t$ is the *stochastic* ('coloured' noise) error component which has the one-sided moving-average (MA) representation:

$$e_t = \sum_{k=0}^{q} \beta_k u_{t-k} = B(z)u_t , \qquad \qquad \text{... (6.3)}$$

in which $u_t$ is white noise with zero mean ($E[u_t] = 0$) and with statistical properties: $E[u_t u_s] = \sigma_u^2 \delta_{ts}$ ($\delta_{ts}$ = Kronecker delta function). We also assume that $u_t$ is mutually uncorrelated with $v_t$, $E[u_t^* v_{t+\lambda}] = 0$.

Substituting equations (6.2) and (6.3) into (6.1), we obtain the combined autoregressive moving-average (ARMA) model:

$$x_t = -\sum_{j=1}^{p} \alpha_j x_{t-j} + \sum_{k=0}^{q} \beta_k u_{t-k} \qquad \qquad \text{... (6.4)}$$
$$= -A_1(z)x_t + B(z)u_t$$

that is,

$$A(z)x_t = B(z)u_t \qquad \qquad \text{... (6.5)}$$

where $A(z) = 1 + A_1(z)$. Equation (6.5) is equivalent to equation (3.22), and may be written in moving-average form as:

$$x_t = \sum_{n=0}^{M-1} g_n u_{t-n} = G(z)u_t \qquad \dots (6.6)$$

where the moving-average parameters $g_n$ are given by equation (3.25) as:

$$g_n = [\beta_n - \sum_{m=1}^{p} g_{n-m}\alpha_m]/\alpha_0 \qquad \dots (6.7)$$

in which $\alpha_0 = \beta_0 = 1$ and $g_{n-m} = 0 \, (n < m)$.

## 6.2.2    Parametric Spectral Density Estimates

It follows from equation (3.23) that equation (6.5) can be written as:

$$G(z) = \frac{X(z)}{U(z)} = \frac{B(z)}{A(z)} = \frac{\sum_{n=0}^{q} \beta_n z^{-n}}{\sum_{m=0}^{p} \alpha_m z^{-m}} \qquad \dots (6.8)$$

Using the usual definition for the power spectral density, $S_{XX}(f)$, we have:

$$\begin{aligned} S_{XX}(f) &= E[X^*(f)X(f)] \\ &= E[\{U^*(f)G^*(f)\}\{U(f)G(f)\}] \qquad \dots (6.9) \\ &= E[U^*(f)U(f)]|G(f)|^2 = \frac{\sigma_u^2|B(f)|^2}{|A(f)|^2} \end{aligned}$$

where $A(f) = 1 + \sum_{k=1}^{p}\alpha_k e^{-j2\pi fk\Delta}$ and $B(f) = 1 + \sum_{k=1}^{q}\beta_k e^{-j2\pi fk\Delta}$ $(z = e^{j\omega\Delta})$.

Equation (6.9) forms the basis for the parametric spectral estimation procedures in what follows. However, it contains *three* unknowns ($\sigma_u^2$, $A(f)$ and $B(f)$) which need to be determined. This is done using the procedures given in section 4.3.2 for the derivation of the Wiener-Hopf convolution equation (4.17), as outlined in section 6.3. The alternative parametric models that can be used in practice are discussed in the following section.

## 6.2.3    Alternative Parametric Models

We see from the above equations that there are three possible types of linear model that can be employed in practice to determine the parametric spectral density estimates:

1) a $p$th order *autoregressive* model, AR($p$); that is, $q = 0$;
2) a $q$th order *moving average* model, MA($q$); that is, $p = 0$;
3) a $(p, q)$th order *autoregressive moving-average* model, ARMA($p$, $q$).

Consider the general case (3) of an ARMA($p$, $q$) model with parameters $\alpha_j$ and $\beta_k$. Using equation (6.4), the even sided autocorrelation function of $M$ lags is defined by:

$$R_{xx}(m) = E[x_t^* x_{t+k}] \qquad\qquad -M \le m \le M$$

$$= -\sum_{j=1}^{p} \alpha_j E[x_t^* x_{t+m-j}] + \sum_{k=0}^{q} \beta_k E[x_t^* u_{t+m-k}]$$

that is,

$$R_{xx}(m) = -\sum_{j=1}^{p} \alpha_j R_{xx}(m-j) + \sum_{k=0}^{q} \beta_k R_{xu}(m-k) \qquad\qquad \text{... (6.10)}$$

The cross-correlation term, $R_{xu}(m-k)$, can be rewritten as (cf. equation (4.12)):

$$R_{xu}(m) = R_{ux}(-m)$$

$$= \sum_{r=0}^{M} g_r R_{uu}(-m-r) = \sigma_u^2 g_m \qquad\qquad \text{... (6.11)}$$

Substituting (6.11) in (6.10) yields:

$$R_{xx}(m) = \begin{cases} -\displaystyle\sum_{j=1}^{p} \alpha_j R_{xx}(m-j) & |m| > q \text{ if } p > q \\[2em] -\displaystyle\sum_{j=1}^{p} \alpha_j R_{xx}(m-j) + \sigma_u^2 \displaystyle\sum_{k=0}^{q-m} g_k \beta_{m+k} & 0 \le |m| \le q \end{cases} \qquad \text{... (6.12)}$$

which, since $g_k$ is a function of the ARMA parameters (equation (6.7)), results in a set of nonlinear equations.

For an AR process ($q = 0$), equation (6.12) simplifies to:

$$R_{xx}(m) = \begin{cases} -\displaystyle\sum_{j=1}^{p} \alpha_j R_{xx}(m-j) + \sigma_u^2 & m = 0 \\[2em] -\displaystyle\sum_{j=1}^{p} \alpha_j R_{xx}(m-j) & |m| > 0 \end{cases} \qquad \text{... (6.13)}$$

which is a linear relationship between the autocorrelation function, $R_{xx}(k)$, and the AR parameters ($\alpha_m$).

For a MA process ($p = 0$), equation (6.12) simplifies to:

$$R_{xx}(k) = \begin{cases} \sigma_u^2 \sum_{n=0}^{q-k} \beta_n \beta_{k+n} & 0 \le |k| \le q \\ 0 & |k| > q \end{cases} \qquad \text{... (6.14)}$$

since the moving-average parameters $g_n = \beta_n$.

Of the three types of model, the *autoregressive* model, AR($p$), is the most widely used for three reasons:

*firstly*, it gives the best frequency resolution, particularly for spectra with narrow peaks (resonances),

*secondly*, its resulting set of linear equations are relatively simple to solve for the AR parameter estimates, and

*thirdly*, Wold's decomposition theorem states that an ARMA or MA process may be uniquely described by an AR process of sufficiently high order or, alternatively, an ARMA or AR process may be uniquely described by an MA process of sufficiently high order.

The last point means that the *linear estimation* problem involves selecting a model with a minimum number of parameters, and this is usually an AR model. Criteria for determining the optimum *model order* are discussed section 6.4.

### 6.3        OPTIMUM WIENER FILTERS

### 6.3.1      Wiener-Hopf Equation

In section 4.3.2, we derived the Wiener-Hopf equation to obtain the 'optimum' transfer (or weighting) function using spectral methods, and we apply the same principles here. From equation (6.3) and (6.4), the AR model is:

$$e_t = u_t = x_t - v_t$$
$$= x_t + \sum_{j=1}^{p} \alpha_j x_{t-j} \qquad \text{... (6.15)}$$

We require the variance of the residual error ($\sigma_u^2$), or *residual error power*,

$$P = E[u_t^* u_t] = \sigma_u^2$$
$$= E[\{x_t^* - v_t^*\}\{x_t - v_t\}] \qquad \text{... (6.16)}$$
$$= \underbrace{E[x_t^* x_t]}_{(1)} - \underbrace{E[x_t^* v_t]}_{(2)} - \underbrace{E[v_t^* x_t]}_{(3)} + \underbrace{E[v_t^* v_t]}_{(4)}$$

to be a minimum. Considering each term on the right hand side of (6.16) in turn we have:

(1) $E[x_t^* x_t] = R_{xx}(0)$

(2) $E[x_t^* v_t] = -\sum_{j=1}^{p} \alpha_j E[x_t^* x_{t-j}] = -\sum_{j=1}^{p} \alpha_j R_{xx}(-j) = -\sum_{j=1}^{p} \alpha_j R_{xx}^*(j) = \mathbf{R}_x^H \mathbf{a}$

(3) $E[v_t^* x_t] = -\sum_{j=1}^{p} \alpha_j^* E[x_{t-j}^* x_t] = -\sum_{j=1}^{p} \alpha_j^* R_{xx}(j) = \mathbf{a}^H \mathbf{R}_x$

(4) $E[v_t^* v_t] = \sum_{j=1}^{p} \sum_{k=1}^{p} \alpha_j^* \alpha_k E[x_{t-j}^* x_{t-k}] = \sum_{j=1}^{p} \sum_{k=1}^{p} \alpha_j^* \alpha_k R_{xx}(j-k) = \mathbf{a}^H \mathbf{R}_{xx} \mathbf{a}$

where the superscript (H) denotes the Hermitian transposition and the $(1 \times p)$ vector $\mathbf{a}^H = [\alpha_1, \ \alpha_2, \ \cdots, \ \alpha_p]$.

Substitution into equation (6.16) yields:

$$P = R_{xx}(0) + \mathbf{R}_x^H \mathbf{a} + \mathbf{a}^H \mathbf{R}_x + \mathbf{a}^H \mathbf{R}_{xx} \mathbf{a} \qquad \ldots (6.17)$$

The error power will be a minimum when the derivative of this equation with respect to the vector $\mathbf{a}$ or $\mathbf{a}^H$ (since $\mathbf{R}_{xx}$ is Hermitian) is zero; that is, when

$$\frac{\partial P}{\partial \mathbf{a}^H} = \mathbf{R}_x + \mathbf{R}_{xx} \mathbf{a}_o = 0$$

or

$$\mathbf{R}_{xx} \mathbf{a}_o = -\mathbf{R}_x \qquad \ldots (6.18)$$

where the subscript (o) denotes the optimum vector satisfying this condition. Equation (6.18) is the Wiener-Hopf integral equation (cf. equation (4.17)), and may be written in expanded matrix form as:

$$\begin{bmatrix} R_{xx}(0) & R_{xx}(-1) & \cdots & R_{xx}(1-p) \\ R_{xx}(1) & R_{xx}(0) & \cdots & R_{xx}(2-p) \\ \vdots & \vdots & \ddots & \vdots \\ R_{xx}(p-1) & R_{xx}(p-2) & \cdots & R_{xx}(0) \end{bmatrix} \begin{bmatrix} \alpha_1 \\ \alpha_2 \\ \vdots \\ \alpha_p \end{bmatrix} = - \begin{bmatrix} R_{xx}(1) \\ R_{xx}(2) \\ \vdots \\ R_{xx}(p) \end{bmatrix} \qquad \ldots (6.19)$$

## 6.3.2 Minimum Error Power

Substitution of equation (6.18) into (6.17) yields the minimum error power as:

$$P_{min} = R_{xx}(0) - \mathbf{R}_x^H \mathbf{a}_o$$

$$= R_{xx}(0) + \sum_{j=1}^{p} \alpha_j R_{xx}^*(j) = \sum_{j=0}^{p} \alpha_j R_{xx}^*(j) \qquad \ldots (6.20)$$

$$= \mathbf{R}_x^H \mathbf{a}$$

where the order of the $\mathbf{R}_x$ and **a** vectors is now $(p+1) \times 1$, and $\alpha_0 = 1$.

It is important to note that equation (6.20) is based on the orthogonality principle $E[e_t^* v_t] = E[x_t^* v_t] - E[v_t^* v_t] = 0$; that is, the predicted output $v_t$ and the error $e_t$ $(= u_t)$ are uncorrelated (independent), and follows from our assumption that $u_t$ is white noise. This implies that, in terms of the original variables, the minimum error power, equation (6.16), is:

$$P_{\min} = E[e_t^* e_t] = E[e_t^* \{x_t - v_t\}] = E[e_t^* x_t]$$
$$= E[\{x_t^* - v_t^*\} x_t] = E[x_t^* x_t] - E[v_t^* v_t] \qquad \ldots (6.21)$$

from which equation (6.20) can be derived directly. We note in passing that this equation satisfies the Pythagorean theorem:

$$\left\{\sqrt{E[x_t^* x_t]}\right\}^2 = \left\{\sqrt{E[e_t^* e_t]}\right\}^2 + \left\{\sqrt{E[v_t^* v_t]}\right\}^2 \qquad \ldots (6.22)$$

and we will use this property in our discussion on prediction error filters in the next section.

### 6.3.3    Prediction Error Filters

A major advantage of the parametric method is that the 'optimum' Wiener filter can be designed to predict the sample value of the stationary time series $\{x_t\}$ on a *d-step ahead* basis (cf. minimum variance controllers, Appendix B5).

Using the above orthogonality principle, the error power:

$$P = E[e_{t+d}^* e_{t+d}] = E[\{x_{t+d}^* - v_t^*\}\{x_{t+d} - v_t\}] \qquad \ldots (6.23)$$

between the predicted value $x_{t+d}$ and the known time series $\{x_t\}$ will be a minimum when:

$$E[\{x_{t+d} - v_t\} x_{t-m}] \qquad\qquad m = 0,1,2,\ldots,M$$
$$= E[\{x_{t+d} - (\gamma_0 x_t + \gamma_1 x_{t-1} + \ldots + \gamma_M x_{t-M})\} x_{t-m}] \qquad \ldots (6.24)$$
$$= 0$$

that is, when

$$R_{xx}(m+d) = \gamma_0 R_{xx}(m) + \gamma_1 R_{xx}(m-1) + \ldots + \gamma_M R_{xx}(m-M)$$
$$= \sum_{k=0}^{M} \gamma_k R_{xx}(m-k) \qquad \ldots (6.25)$$

where the relationship between the coefficients $(\gamma_k)$ of the prediction operator and the previous AR parameters is defined below.

In equation (6.23), we defined the prediction error time series by

$$e_{t+d} = x_{t+d} - (\gamma_0 x_t + \gamma_1 x_{t-1} + \ldots + \gamma_M x_{t-M}) \qquad \ldots (6.26)$$

and it follows that, if we define the prediction error operator by:

$$[\alpha_0, \alpha_1, \ldots, \alpha_{M+d}] = [1, 0, \ldots, 0, -\gamma_0, -\gamma_1, \ldots, -\gamma_M] \qquad \ldots (6.27)$$

then

$$e_{t+d} = \alpha_0 x_t + \alpha_1 x_{t-1} + \ldots + \alpha_{M+d} x_{t-M-d} . \qquad \ldots (6.28)$$

In equation (6.28), we have performed a *backwards time shift* operation, as shown in Figure 6.2. The minimum prediction error power in terms of the prediction error operator is given by:

$$P_{\min} = \alpha_0 R_{xx}(0) + \alpha_1 R_{xx}(1) + \ldots + \alpha_{M+d} R_{xx}(M+d) \qquad \ldots (6.29)$$

Equation (6.29) is a more general form of equation (6.20), in which $p = M$ and $d = 0$, and we make the following observations:
1) if we normalise the minimum prediction error power with respect to $R_{xx}(0)$, then the resulting *normalised prediction error power* (say $Q_{\min}$) will lie between *zero* (perfect prediction) and *unity* (completely unpredictable) for a weakly stationary random processes; and
2) when the term 'prediction-error filter' is used without explicit reference to the prediction distance, $d$, then it is usually assumed that it is a *one-step* ahead prediction error filter ($d = 1$).

Since *one-step ahead* prediction-error filters are most commonly used in practice, we will confine the following analysis to this the latter class of filters.

### 6.3.4    Prediction-Error Filter Equations

From the above analysis, we see that the autocorrelation between $x_t$ and the one-step ahead prediction $x_{t+1}$ is:

$$R_{xx}(m+1) = E[x_t^* x_{t+1+m}] \qquad m = 0, 1, \ldots, M-1 \qquad \ldots (6.30)$$

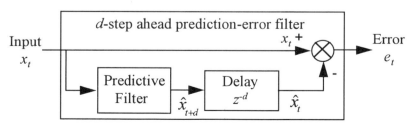

**Figure 6.2** Forward linear prediction-error filter

where we have confined the number lags to $M$ lags. Augmenting equations (6.18) or (6.19) by (6.20) and redefining $\mathbf{R}_{xx}$ to be of order $(M+1)\times(M+1)$, then we obtain (with $p = M$):

$$\mathbf{R}_{xx}\mathbf{a} = P_M\mathbf{i} \qquad \text{... (6.31}a\text{)}$$

which may be written in expanded matrix form as:

$$
\begin{bmatrix}
R_{xx}(0) & R_{xx}(-1) & \cdots & R_{xx}(-M) \\
R_{xx}(1) & R_{xx}(0) & \cdots & R_{xx}(1-M) \\
\vdots & \vdots & \ddots & \vdots \\
R_{xx}(M) & R_{xx}(M-1) & \cdots & R_{xx}(0)
\end{bmatrix}
\begin{bmatrix}
1 \\
\alpha_2 \\
\vdots \\
\alpha_M
\end{bmatrix}
=
\begin{bmatrix}
P_M \\
0 \\
\vdots \\
0
\end{bmatrix}
\qquad \text{... (6.31}b\text{)}
$$

or equivalently,

$$\sum_{k=0}^{M}\alpha_k R_{xx}(m-k) = \begin{cases} P_M & m = 0 \\ 0 & m = 1,2,\ldots,M \end{cases} \qquad \text{... (6.31}c\text{)}$$

where $P_M$ denotes the corresponding *minimum error power* of the $M$th order prediction-error filter, and is given by equation (6.20) or (6.31$c$) as:

$$P_M = \sum_{k=0}^{M}\alpha_k R_{xx}(k) \qquad \text{... (6.32)}$$

### 6.3.5     The Yule-Walker Method for AR Parameter Estimation

The Yule-Walker method takes advantage of the Toeplitz form of the autocorrelation matrix, $\mathbf{R}_{xx}$, in computing the AR parameter estimates. Only the essential features of the method are described here.

The steps for solving equations (6.31) essentially involve:

1) computing the autocorrelation estimates, $\hat{R}_{xx}(k)$, from the observed data using a *biased* form of the autocorrelation function (equation (3.79)):

$$R_{xx}(m) = \frac{1}{N}\sum_{k=0}^{N-|m|}x^*(k)x(m+k) \qquad \text{... (6.33)}$$

    ($m = 0,M$) to ensure that the autocorrelation matrix is positive semi-definite and that the resulting AR model is stable, and

2) inverting the autocorrelation matrix, equations (6.31), using the Levinson algorithm (Appendix A6) to obtain the AR parameters by minimisation of the forward and backward errors.

The corresponding power spectral density estimate is then computed from:

$$S_{XX}^{YW}(f) = \frac{P_M}{2B\left|1 + \sum_{m=1}^{M} \hat{\alpha}_m e^{-j2\pi f m\Delta}\right|^2} \qquad \text{... (6.34)}$$

where $\hat{\alpha}_m$ are the AR parameter estimates obtained from the recursion, the estimated minimum error power for the $M$th order filter is,

$$\hat{P}_M = \hat{\sigma}_x^2 \prod_{m=1}^{M} [1 - |\hat{\alpha}_m|^2] \qquad \text{... (6.35)}$$

$B$ is the frequency bandwidth $(-B \leq f \leq B)$; and the sample interval $(\Delta)$ is normally equivalent to the Nyquist rate: $\Delta = 1/2B$.

A drawback of the Yule-Walker method is that it is limited by the *a priori* information contained in the autocorrelation estimates, $\hat{R}_{xx}(k)$ $(-M \leq k \leq M)$, and assumes that, consistent with conventional spectral estimation, the data outside the window of available information are zero.

## 6.3.6    The Maximum-Entropy Spectral Estimator

To overcome the limitation of parameter (hence spectral) estimates based on finite autocorrelation information, more recent parametric spectral methods have utilised the *entropy* (a measure of the information content) of the process, which is maximised to obtain best estimates of the autocorrelation function, $\hat{R}_{xx}(m), 0 \leq m \leq M$, outside its maximum lag range $(m > M)$ such that the autocorrelation matrix, $\mathbf{R}_{xx}$, is positive semidefinite.

For a process with a Gaussian probability density distribution, the total entropy is given by:

$$H = \tfrac{1}{2}\ln[\det(\mathbf{R}_{xx})] \qquad \text{... (6.36)}$$

However, for a process of infinite duration, the entropy $H \to \infty$, and so the asymptotically stable *entropy rate*:

$$h = \lim_{M \to \infty} \frac{H}{M+1} = \lim_{M \to \infty}\left[\tfrac{1}{2}\ln\{\det(\mathbf{R}_{xx})\}^{1/(M+1)}\right] \qquad \text{... (6.37)}$$

is used.

Now it can be shown that the limiting form of the autocorrelation matrix is related to the spectral density $S_{XX}(f)$ by the relationship:

$$\lim_{M \to \infty} [\det(\mathbf{R}_{xx})]^{1/(M+1)} = 2B \exp\left\{ \frac{1}{2B} \int_{-B}^{B} \ln[S_{XX}(f)]df \right\} \qquad \text{... (6.38)}$$

Substituting (6.38) into (6.37) yields the entropy rate of the process as:

$$h = \tfrac{1}{2}\ln(2B) + \frac{1}{4B} \int_{-B}^{B} \ln[S_{XX}(f)]df \qquad \text{... (6.39)}$$

Note that, although this relation assumes that the process has a Gaussian probability distribution, it holds for any stationary times series.

The estimation problem involves determining the spectral density estimates, $\hat{S}_{XX}(f)$, such that the entropy rate $h$, equation (6.39), is both *stationary* with respect to the unknown autocorrelation values *outside* the range $0 \le |m| \le |M|$, and *consistent* with respect to the known autocorrelation values *inside* the range $0 \le |m| \le |M|$.

Recall that the observed power spectral density, $S_{XX}(f)$, is related to its autocorrelation function, $R_{xx}(\tau)$, by (cf. equation (2.53)):

$$S_{XX}(f) = \text{FT}[R_{xx}(m)] = \Delta \sum_{m=-\infty}^{\infty} R_{xx}(m)e^{-j2\pi f\Delta} \qquad \text{... (6.40)}$$

where $\Delta$ is the sample interval and $\tau = m\Delta$. Substitution in (6.39) yields:

$$h = \frac{1}{4B} \int_{-B}^{B} \ln\left[ \sum_{m=-\infty}^{\infty} R_{xx}(m)e^{-j2\pi f\Delta} \right] df \qquad \text{... (6.41)}$$

Minimisation of this equation for the entropy rate $h$ with respect to $R_{xx}(m)$ yields the following condition:

$$\frac{\partial h}{\partial R_{xx}(m)} = \int_{-B}^{B} \frac{e^{-j2\pi f\Delta}}{\hat{S}_{XX}(f)} df = 0 \qquad |m| \ge M+1 \qquad \text{... (6.42)}$$

which states that the unknown autocorrelation values add no new information or entropy to the process. Note that $\hat{S}_{XX}(f)$ is now the spectral density estimate constrained by equation (6.42), and may be expressed as a truncated Fourier series by:

$$\frac{1}{\hat{S}_{XX}(f)} = \sum_{n=-M}^{M} c_n e^{-j2\pi f\Delta} \qquad \text{... (6.43)}$$

where $c_{-n} = c_n^*$ to ensure that $\hat{S}_{XX}(f)$ is real symmetric.

The RHS of equation (6.43) can be written in z-operator form as:

$$\sum_{n=-M}^{M} c_n z^{-n} = \frac{2B}{P_M} A_M(z) A_M^*(z) = \frac{2B}{P_M} |A_M(z)|^2 \qquad \text{... (6.44)}$$

where $A_M(z)$ denotes the $M$th order AR model, equation (6.31), and the coefficient $2B/P_M$ is the (static gain) value at $n = 0$.

Substitution of (6.44) in equation (6.43) yields:

$$\hat{S}_{XX}^{ME}(f) = \frac{P_M}{2B \left| 1 + \sum_{m=1}^{M} \hat{\alpha}_m e^{-j2\pi f\Delta} \right|^2} \qquad \text{... (6.45)}$$

This equation is the *maximum-entropy spectral density estimate* of the process, and is identical in form to that obtained by least squares estimation of the AR parameters $(\hat{\alpha}_m)$ from the observed time series $\{x_t\}$. Thus, the principle of maximum entropy and the representation of a process by an AR model are equivalent. The Burg method discussed in the following section is based on this equivalence relationship, but it should be noted that more recent studies involving general formulations of the maximum entropy method (MEM) result in a set of nonlinear equations, and this equivalence only strictly applies when the autocorrelation estimates, $R_{xx}(m)$, are known exactly.

### 6.3.7    AR Parameter Estimation – Burg Method

The Burg method is a recursive least squares lattice method which attempts to maximise the *entropy* (or information content) of the process by minimising the forward and backward errors in the prediction error (lattice) filters with the restriction that the AR parameters satisfy the Levinson-Durbin recursion.

Using equations (6.1) to (6.3), we begin by passing through the data in the *forward* direction and obtaining an estimate of the *forward error* at time $t$ as:

$$e_t^f = x_t - \hat{x}_t = x_t - \sum_{m=1}^{p} \alpha_m x_{t-m} \qquad \text{... (6.46)}$$

In a similar manner, we compute the *backward error* of the prediction estimate in a *backward* direction as:

$$e_t^b = x_{t-p} - \hat{x}_{t-p} = x_{t-p} - \sum_{m=1}^{p} \alpha_m^* x_{t+m-p} \qquad \text{... (6.47)}$$

where $\alpha_m^*$ is the complex conjugate value computed in the backward direction.

We define the $p$th order *error power* by:

$$P_p = \tfrac{1}{2}\left\{P_p^f + P_p^b\right\} = \tfrac{1}{2}\left\{E\left[\left|e_t^f\right|^2 + \left|e_t^b\right|^2\right]\right\}$$

$$= \tfrac{1}{2}\sum_{t=p}^{N-1}\left[\left|x_t - \sum_{m=1}^{p}\alpha_m x_{t-p}\right|^2 + \left|x_{t-p} - \sum_{m=1}^{p}\alpha_m^* x_{t+m-p}\right|^2\right] \qquad \ldots (6.48)$$

and we then minimise this error power with respect to the AR parameters in turn to obtain a set of equations which satisfy the Levinson recursion:

$$\alpha_m^p = \alpha_m^{p-1} + \rho_p \alpha_{p-m}^{(p-1)*} \qquad \ldots (6.49)$$

where $\rho_p$ is the $p$th *reflection coefficient*, and $\alpha_m^p$ and $\alpha_m^{p-1}$ are the $m$th AR parameter estimates computed in the $p$th and the previous $((p-1)$th) steps respectively.

It can be shown that the $p$th order prediction error power, equation (6.38), is related to the $(p-1)$th order prediction error power calculated in the previous step by:

$$P_p = P_{p-1}\left(1 - \left|\rho_p\right|^2\right) \qquad \ldots (6.50)$$

The computational procedure is performed in the following steps.

1) We calculate the prediction-error power, $P_0$, for an AR model of zero order $(p = 0)$, which equals $\hat{R}_{xx}(0)$; that is, $P_0 = \hat{R}_{xx}(0) = \hat{\sigma}_x^2$.
2) We then set $p = 1$ and calculate the prediction-error power, $P_1$, using equation (6.48), and then determine the value of the reflection coefficient $\rho_1$ at which $P_1$ is a minimum; that is, by solving $\partial P_1 / \partial \rho_1 = 0$.
3) We proceed in a similar manner for higher orders of $m$ and obtain the following general expression (known as Burg's formula) for the $m$th reflection coefficient:

$$\rho_m = \frac{-2\sum_{t=m+1}^{N}e_{t,m-1}^f e_{t-1,m-1}^{b*}}{\sum_{t=m+1}^{N}\left[\left|e_{t,m-1}^f\right|^2 + \left|e_{t-1,m-1}^{b*}\right|^2\right]} \qquad m = 1, 2, \ldots, p \qquad \ldots (6.51)$$

where the forward and backward prediction error terms are calculated using equation (6.48), and may be computed by the recursive relation after Andersen (1974):

$$\hat{P}_m = \left[1 - |\rho_m|^2\right]\hat{P}_{m-1} - \left|e^f_{m-1,m-1}\right|^2 - \left|e^b_{m-2,m-2}\right|^2 \qquad \text{... (6.52)}$$

4) We then compute the prediction error filter parameters using equation (6.49) and repeat steps 1 to 4 for the desired filter order ($p$).
5) Finally, we compute the power spectral density estimate from:

$$S^B_{XX}(f) = \frac{\hat{E}_p}{2B\left|1 + \sum_{m=1}^{p}\hat{\alpha}_m e^{-j2\pi f m\Delta}\right|^2} \qquad \text{... (6.53)}$$

Advantages of the Burg method are:
1) high frequency resolution;
2) the resulting AR prediction error filter is *stable*;
3) it is computationally efficient.
Some disadvantages, however, are that:
1) it exhibits a phenomenon known as *spectral line splitting* at high signal to noise ratios (SNRs), in which the Burg method gives *two* closely spaced peaks when the true spectrum has only a *single* sharp peak;
2) it can give spurious peaks in the spectra for higher order models;
3) it exhibits a *frequency shift* from the true frequency which is sensitive to the *initial phase* of the periodic signal, especially for *short* data records.
Various modifications have been proposed to overcome these deficiencies of the Burg algorithm, and these basically involve the inclusion of a *weighting sequence* (data window) in the computation of the prediction error power, equation (6.48), which is replaced by:

$$P_p = \tfrac{1}{2}\left[P^f_p + P^b_p\right] = \tfrac{1}{2}\left\{E\left[w_{t,p}\left\{\left|e^f_t\right|^2 + \left|e^b_t\right|^2\right\}\right]\right\} \qquad \text{... (6.54)}$$

Minimisation of this equation results in the weighted reflection coefficient estimates:

$$\rho_m = \frac{-2\sum_{t=m+1}^{N} w_{t,m-1} e^f_{t,m-1} e^{b*}_{t-1,m-1}}{\sum_{t=m+1}^{N} w_{t,m-1}\left[\left|e^f_{t,m-1}\right|^2 + \left|e^{b*}_{t-1,m-1}\right|^2\right]} \qquad m = 1, 2, ..., p \qquad \text{... (6.55)}$$

The application of Hamming and other data window functions proposed by various researchers (Proakis *et al.* 1992) has proved effective in reducing the problems of line splitting, spurious peaks and frequency bias.

### 6.3.8 AR Parameter Estimation – Least Squares Method

The Burg method described in the previous section is a least squares lattice algorithm with predictor coefficients constrained by the Levinson recursion. The following *unconstrained* least squares method for determining the AR parameters relaxes this constraint, and obviates some the disadvantages of the Burg algorithm.

We compute the error power from the forward and backward errors as in equation (6.48):

$$P_p = \tfrac{1}{2}\left\{P_p^f + P_p^b\right\} = \tfrac{1}{2}\left\{E\left[\left|e_t^f\right|^2 + \left|e_t^b\right|^2\right]\right\}$$

$$= \tfrac{1}{2}\sum_{t=p}^{N-1}\left[\left|x_t - \sum_{m=1}^{p}\alpha_m x_{t-m}\right|^2 + \left|x_{t-p} - \sum_{m=1}^{p}\alpha_m^* x_{t+m-p}\right|^2\right] \qquad \text{... (6.56)}$$

and obtain the matrix equation:

$$P_p = \left(\mathbf{x_p} - \mathbf{X_f}\mathbf{a}\right)^T\left(\mathbf{x_p} - \mathbf{X_f}\mathbf{a}\right) + \left(\mathbf{x_1} - \mathbf{X_b}\mathbf{a}\right)^T\left(\mathbf{x_1} - \mathbf{X_b}\mathbf{a}\right) \qquad \text{... (6.57)}$$

where

$$\mathbf{x_p^T} = [x_{p+1}, x_{p+2}, \ldots, x_{N-1}]$$
$$\mathbf{x_1^T} = [x_1, x_2, \ldots, x_{N-p}] \qquad \text{... (6.58)}$$

$$\mathbf{X_f} = \begin{bmatrix} x_p & x_{p-1} & \cdots & x_1 \\ \vdots & & & \vdots \\ x_{2p} & & & x_p \\ \vdots & & & \vdots \\ x_{N-1} & x_{N-2} & \cdots & x_{N-p} \end{bmatrix} \qquad \mathbf{X_b} = \begin{bmatrix} x_1 & x_2 & \cdots & x_p \\ \vdots & & & \vdots \\ x_p & & & x_{2p} \\ \vdots & & & \vdots \\ x_{N-p} & x_{N-p+1} & \cdots & x_{N-1} \end{bmatrix}$$

and the $(p \times 1)$ column vector of AR parameters, $\mathbf{a}$, is defined as previously. Equation (6.57) is minimised with respect to $\mathbf{a}$ to yield:

$$\left(\mathbf{X_f^T X_f} + \mathbf{X_b^T X_b}\right)\mathbf{a} = \mathbf{X_f^T x_p} + \mathbf{X_b^T x_1}$$
$$\mathbf{R_{xx}}(p)\mathbf{a} = \mathbf{R_x}(p) \qquad \text{... (6.59)}$$

where the elements of the $(p \times p)$ correlation matrix, $\mathbf{R_{xx}}(p)$, are given by

$$R_{xx}(j,k) = \sum_{t=p}^{N-1} x_{t-j} x_{t-k} + \sum_{t=p}^{N-1} x_{t-p+j} x_{t-p+k} \qquad j,k = 1,2,\ldots,p \qquad \text{... (6.60)}$$

and those of the $(p \times 1)$ column vector, $\mathbf{R}_x(p)$, are given by:

$$R_{xx}(j) = \sum_{t=p}^{N-1} x_{t-j} x_t + \sum_{t=p}^{N-1} x_{t-p+j} x_{t-p} \qquad j = 1, 2, \ldots, p \qquad \ldots (6.61)$$

We note that the correlation matrix, $\mathbf{R}_{xx}(p)$, is a symmetrical matrix but not necessarily Toeplitz. However, it approaches the Toeplitz form as $N \to \infty$. Hence, equation (6.59) cannot be solved using the Levinson recursion, and other recursive schemes such as the lattice algorithm of Marple (1987) should be used.

The least squares method has the advantage that is less sensitive to initial phase than Burg's method, but the resolution frequency of the prediciton error filter is broader than Burg's method for the equivalent number of parameters.

There are a number of other parametric spectral methods reported in the literature, and the reader is referred to Haykin (1979), Marple (1987) and Proakis *et al.* (1992) for further information. We conclude this review with a brief discussion on parameter adaptive methods in the following section.

### 6.3.9    AR Parameter Estimation – Adaptive Methods

The above methods utilise block data which are assumed to be weakly stationary; that is, their first and second order statistics are independent of time. Consistent with trends in modern control theory, parameter adaptive estimation methods are also employed to investigate *nonstationary* data, or where the dominant frequency components change with time.

In principle, weakly nonstationary processes can be investigated by segmenting the total data record into shorter data records with approximately stationary properties. However, the problems inherent with windowing and short data records are amplified in this approach. A more elegant approach is to model the data with time varying coefficients using parameter adaptive methods.

By way of an example, let us consider the first order AR process:

$$x_t = a_1 x_{t-1} + e_t \qquad \ldots (6.62)$$

The standard least mean square approach minimises the mean square error:

$$\varepsilon = \sum_t e_t^2 = \sum_t (x_t - a_1 x_{t-1})^2 \qquad \ldots (6.63)$$

with respect to the parameter $a_1$; that is, $\partial \varepsilon / \partial a_1 = 0$. However, since $a_1$ is time-dependent, $\partial \varepsilon / \partial a_1$ will also deviate from zero with time, and the adaptive algorithm is designed to minimise this departure. This is done by requiring

$$\varepsilon_t = e_t^2 = (x_t - a_1 x_{t-1})^2 \qquad \dots (6.64)$$

to be a minimum at each time step; that is,

$$\frac{\partial \varepsilon_t}{\partial a_1} = \nabla[e_t^2] = -2(x_t - a_1 x_{t-1})x_{t-1} = -2e_t x_{t-1} \to 0 \qquad \dots (6.65)$$

Equation (6.65) suggests that we can iteratively adjust the parameter $a_1$ such that:

$$a_1 \to a_1 + \mu e_t x_{t-1} \to a_1 + \eta \nabla[e_t^2] \qquad \dots (6.66)$$

is minimised by the method of steepest descent, where $\mu \, (= -2\eta)$ is a convergence factor that adjusts the amount of correction we make. The choice of $\mu$ affects both the time constant and stability of the adaptive algorithm, and it can be shown that its stability bounds must satisfy $2/\lambda_{max} > \mu > 0$, where $\lambda_{max}$ is the maximum eigenvalue of the autocorrelation matrix. As a general rule, the faster the adaption the poorer the performance, since the adaptive filter will also try to compensate for (broadband) additive noise.

Extending the concepts underlying equation (6.66) to the AR parameter vector, **a**, the predicted parameters at the $(t+1)$th time step is given by:

$$\mathbf{a}_{t+1} = \mathbf{a}_t + \mu e_t \mathbf{x}_{t-1} \qquad \dots (6.67)$$

where the initial parameter values ($\mathbf{a}_0$) may be determined using any of the above (stationary) methods.

## 6.4      MODEL ORDER CRITERIA

A major difficulty with parametric spectral analysis methods is the selection of the 'optimum' order of the AR prediction error filter (i.e. $p$ or, equivalently, the number of autocorrelation lags $M$). If the order of the AR prediction error filter (PEF) is too small, resonances in the data cannot be resolved, and the spectral estimates will be *biased*. If the order of the PEF is too large, spurious peaks (instabilities) occur in the spectral estimates which result in a large error variance.

A wide range of model order criteria (MOC) have been developed to aid in the selection of an 'optimum' PEF, and these are usually based on either:

1)  a weighted residual error variance, $\sigma_e^2$, or minimum error power, $P_M$; or
2)  the variance of the AR parameter statistics.

Ideally, we would like the 'optimum' PEF to have both *low residual variance* and *low parameter variance*, and some researchers advocate minimising the residual variance first, then minimising the parameter variance. However, the latter approach is more relevant to adaptive methods, and so we will only consider the first type here.

### 6.4.1 Residual Error Variance (REV)

A simple, logical approach is to track the change in the error power ($P_M$) as the number of autocorrelation lags ($M$) is increased; that is, we define a residual error variance (or power) for the $M$th order PEF as:

$$REV(M) = \sigma_e^2(M) = P_M \qquad \text{... (6.68)}$$

Unfortunately, this simple method often does not yield a pronounced minimum and is fairly insensitive to changes in model order as $M$ is increased.

### 6.4.2 Coefficient of Determination (COD)

This is a modified version of the REV criteria, and is defined by:

$$COD(M) = \left(1 - \frac{\sigma_e^2(M)}{\sigma_x^2}\right) \qquad \text{... (6.69)}$$

where $\sigma_x^2$ is the variance of the measurand. This method is similarly insensitive to changes in model order as $M$ is increased, but it can give sufficiently reliable results in some applications.

### 6.4.3 Final Prediction Error (FPE)

The FPE criterion developed by Akaike (1969) for determining the order of AR models is perhaps the most widely known weighted residual error method, and is given by:

$$FPE(M) = \left(\frac{N+M+1}{N-M-1}\right)P_M \qquad \text{... (6.70)}$$

where $N$ is the number of data points. The FPE criterion gives excellent results for processes which can be adequately described by an AR model. However, it is known to *underestimate* the model order with increasing SNRs.

### 6.4.4 Akaike Information Criterion (AIC)

The AIC is an information theoretic criterion derived by Akaike (1974) using maximum likelihood principles, and it gives a compromise between ARMA model complexity and goodness of fit. The AIC in its general form is given by:

$$\text{AIC}(M) = -2\ln(\text{maximum likelihood}) + 2(\text{number of free parameters}).$$
$$\dots (6.71a)$$

For normally distributed ARMA processes, this reduces to:

$$\text{AIC}(p,q) = N\ln(P_{p,q}) + 2(p+q) \qquad \dots (6.71b)$$

and for AR processes ($q = 0$):

$$\text{AIC}(M) = N\ln(P_M) + 2M \qquad \dots (6.71c)$$

From equations (6.70) and (6.71c), it can be shown that, for $N \gg M$:

$$\text{AIC}(M) \approx N\ln\{FPE(M)\}$$

and hence, like the FPE criterion, the AIC tends to *underestimate* the model order with increasing SNRs. To overcome this problem, equation (6.71c) is sometimes modified to:

$$\text{AIC}(M) = N\ln(P_M) + \lambda M \qquad \dots (6.71d)$$

where $\lambda\ (\geq 2)$ is a weighting factor designed to increase the model order and hence model complexity. Akaike also developed a Bayesian information criterion (BIC) based on Bayes' theorem, and obtained

$$\text{BIC}(p,q) = N\ln(P_{p,q}) + 16(p+q)$$
$$= \text{AIC}(p,q) + 14(p+q) \qquad \dots (6.71e)$$

from which it can be seen that $\lambda = 14$.

A further disadvantage of the AIC (and BIC) is that they are statistically inconsistent as $N \to \infty$ (Kashyap 1980).

### 6.4.5    Minimum Description Length (MDL)

Rissanen (1983) proposed an alternative information criterion which *minimises the description length* (MDL), and is defined by:

$$\text{MDL}(M) = N\ln(P_M) + M\ln(N) \qquad \dots (6.72)$$

In contrast to the AIC, the MDL criterion is statistically consistent as $N \to \infty$.

Note that, in applying the above criteria, the mean should be removed from the data. Futhermore, they should be used as a guide only, since extensive experimental results reported in the literature indicate that the above model-order selection criteria do not yield definitive results in every case.

## 6.5 PARAMETRIC CROSS-SPECTRAL ANALYSIS

Ulrych and Jensen (1974) extended the above maximum entropy method (MEM) concepts for power spectra analysis to the computation of the MEM cross-spectra from a knowledge of the MEM power spectra. Their analysis provides an useful bridge between the *non-parametric* bivariate spectral analysis methods discussed in Chapter 4 and the multivariate spectral analysis methods discussed in Chapter 7.

Consider the bivariate time series $\{x_t\}$ and $\{y_t\}$. We define two new time series:

$$v_t = x_t + jy_t \qquad \ldots (6.73a)$$

and

$$w_t = x_t + y_t. \qquad \ldots (6.73b)$$

Taking Fourier transforms, we obtain

$$V(f) = X(f) + jY(f) \qquad \ldots (6.74a)$$

and

$$W(f) = X(f) + Y(f). \qquad \ldots (6.74b)$$

Using the usual definitions, the power spectral density of the derived series $\{v_t\}$ and $\{w_t\}$ are:

$$
\begin{aligned}
S_{VV}(f) &= E[V^*(f)V(f)] \\
&= E[\{X^*(f) - jY^*(f)\}\{X(f) + jY(f)\}] \\
&= E[\{X^*(f)X(f) + Y^*(f)Y(f)\} - j\{Y^*(f)X(f) - X^*(f)Y(f)\}] \\
&= S_{XX}(f) + S_{YY}(f) - 2\,\mathrm{Im}\{E[X^*(f)Y(f)]\}
\end{aligned}
$$

$$\ldots (6.75)$$

and, similarly,

$$
\begin{aligned}
S_{WW}(f) &= E[W^*(f)W(f)] \\
&= E[\{X^*(f) + Y^*(f)\}\{X(f) + Y(f)\}] \\
&= E[\{X^*(f)X(f) + Y^*(f)Y(f)\} + \{Y^*(f)X(f) + X^*(f)Y(f)\}] \\
&= S_{XX}(f) + S_{YY}(f) + 2E[\mathrm{Re}\{X^*(f)Y(f)\}]
\end{aligned}
$$

$$\ldots (6.76)$$

Now the cospectral density and quadspectral density functions of $\{x_t\}$ and $\{y_t\}$ are defined respectively as:

$$C_{XY}(f) = E[\text{Re}\{X^*(f)Y(f)\}]$$
$$= \frac{1}{2}[S_{WW}(f) - S_{XX}(f) - S_{YY}(f)] \qquad \ldots (6.77)$$

and

$$Q_{XY}(f) = E[\text{Im}\{X^*(f)Y(f)\}]$$
$$= \frac{1}{2}[S_{VV}(f) - S_{XX}(f) - S_{YY}(f)] \qquad \ldots (6.78)$$

from which the cross-spectral density is given by:

$$S_{XY}(f) = C_{XY}(f) - jQ_{XY}(f) \qquad \ldots (6.79)$$

A disadvantage of this approach is that, if $\{x_t\}$ and $\{y_t\}$ are contaminated by broadband noise, then, as shown in section 4.6.1, Chapter 4, both the cross-spectral density (hence transfer function) modulus *and* phase estimates will be affected, in contrast to only the modulus estimates computed by DFT methods, equation (4.23).

## 6.6      SIMULATION EXPERIMENTS

### 6.6.1      Computational Procedure and Data Input

The simulation tests discussed below were performed using the computer program MESA#SIM.EXE supplied with the text for IBM/PC compatible computers operating under Windows 3.1 and higher (see Appendix B6).

   The *aim* of this simulation is to demonstrate the effects of increasing data record lengths and AR model order on the parametric spectral (MESA) estimates as compared to those computed by standard discrete Fourier transform (DFT) methods. The simulation experiments are performed using the following computational procedure:

1) $N$ data values of the time series, $\{x_t\}$, are generated using the formula:

$$x_t = \sin(2\pi f_L t) + \sin(2\pi f_H t) + e_t, \qquad \ldots (6.80)$$

where $f_L$ and $f_H$ are the low and high frequencies respectively, and $e_t$ is the random noise term generated using the central limit theorem, and whose amplitude is controlled by the signal to noise ratio (SNR) in the program data input;

2) the AR model order, $M$, is set at a fixed value, but is varied by the program when $N \le M$ to $M = N - 1$;
3) the data record length is doubled for each successive case, and is varied between 10 and 80 (10, 20, 40, 80);
4) the DFT spectral values were computed using a *rectangular* data window for maximum frequency resolution;

The program data input are given in Table 6.1. The sampling rate is 0.1 seconds, which corresponds to a sampling frequency of 5 Hz. The lower and upper frequencies, $f_L$ and $f_H$, are set at 1.9 Hz and 2.1 Hz respectively, and the SNR at 0.1 ($\equiv -20$ dB). The AR model order is nominally 20, and we have chosen to use Andersen's algorithm (Andersen 1974).

### 6.6.2 Experimental Results and Discussion

The results for the four cases tested are shown in Figure 6.3, from which the following observations can be drawn.

1) *Frequency Resolution*: The MESA spectral estimates give much higher frequency resolution than the DFT spectral estimates which, as expected, are unable to resolve the resonant frequencies (shown by spectral lines) for short data lengths, and only give 'acceptable' results when $N \ge 80$. However, the relative magnitudes of the MESA spectral estimates at the resonant frequencies are markedly different, and there is a pronounced 'frequency shift' when $M \cong N$ for short data records ($N = 10, 20$), Figures 6.3a and 6.3b.
1) *Model Order*: the MESA spectral estimates also yield 'spurious' peaks in the cases when $M \cong N$, Figures 6.3a and 6.3b, which indicates that the AR model order is too large for estimating the spectral information contained in the short data records. Akaike's final prediction error (FPE) criterion, equation (6.70), yields an optimum model order of four parameters ($M_{opt} = 4$), Figure 6.4. However, inspection of the resulting MESA spectral estimates shows a dominant peak at 1.9 Hz and a second

Table 6.1 MESA#SIM program - simulation test data input

| | | | |
|---|---|---|---|
| Points/Block | 10 | No. of Frequencies | 200 |
| Total Points | 10 | Method [A=1: H=2] | 1 |
| Time Step (s) | 0.1 | AR Model Order | 20 |
| Noise/Signal Ratio | 0.1 | Optimise Order [N=0: Y=1] | 0 |
| Low Sine Wave Frequency | 1.9 | Order Criterion [1→5] | 4 |
| Low Sine Gain | 1 | Akaike Factor | 1 |
| High Sine Wave Frequency | 2.1 | Residual Noise Filter Order | 0 |
| High Sine Gain | 1 | New Data/Block | 0 |

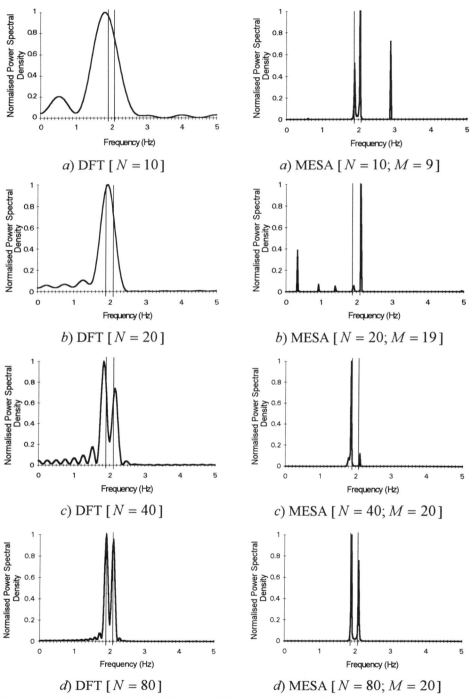

**Figure 6.3** Comparison of DFT and MESA power spectral density estimates for two sinusoids

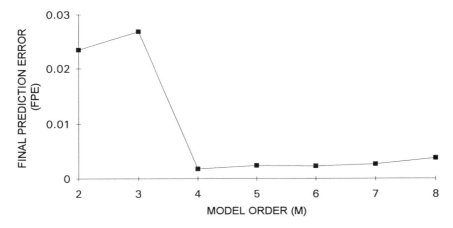

**Figure 6.4** Final prediction error (FPE) criterion for 10 data points

'blip' (5% magnitude) at 2.1 Hz. The second peak grows in magnitude with increase in model order, and the best MESA spectral estimates are obtained for $M = 7$, in which the resonant peaks are most pronounced. However, for $M = 6$, the *second* peak is 55% of the magnitude of the first peak, whereas when $M = 7$, the *first* peak is 85% of the magnitude of the second peak. For $M = 8$ the modulus of the first peak drops to 65%, and when $M = 9$ it drops further to 50% and a third peak (70%) at 2.9 Hz appears. This changeover in magnitude with model order is symptomatic of parametric spectral methods, and the higher frequency components usually become more pronounced as the model order is increased. (The reader should verify these results using the MESA#SIM program.) It should be noted that the FPE criterion, Figure 6.4, tends to infinity ($FPE \rightarrow \infty$) when $N = 9$.

A wide variety of simulation studies can be performed using the MESA#SIM program, and these are explored further in the tutorial exercises.

## REFERENCES

Akaike, H., 1969, Power Spectrum Estimation Through Autoregression Model Fitting, *Ann. Inst. Math.*, **21**, 407-419.

Akaike, H., 1974, A New Look at the Statistical Model Identification, *IEEE Trans. Automatic Control*, **AC-19**, 716-723, December.

Andersen, N. O., 1974, On the Calculation of Filter Coefficients for Maximum Entropy Spectral Analysis, *Geophysics*, **39**, No. 1, Feb., 69-72.

Haykin, S., 1979, *Nonlinear Methods of Spectral Analysis*, S Haykin, ed., Topics in Applied Physics, Springer-Verlag, Berlin.

Kashyap, R. L., 1980, Inconsistency of the AIC Rule for Estimating the Order of Autoregressive Models, *IEEE Trans. Automatic Control*, **AC-25**, 996-998.

Marple, S. L., 1987, *Digital Spectral Analysis with Applications*, Prentice-Hall, Englewood Cliffs, N. J.

Proakis, J. G., Rader, C. M., Ling, F. & Nikias, C. L., 1992, *Advanced Digital Signal Processing*, Maxwell Macmillan International, New York.

Rissanen, J., 1983, A Universal Prior for the Integers and Estimation by Minimum Description Length, *Ann. Stat.*, **11**, 417-431.

Ulrych, T., & Jensen, O., 1974, Cross-spectral Analysis Using Maximum Entropy, *Geophysics*, **39**, No. 3, 353-354, June.

## TUTORIAL EXERCISES

1) Case Study F investigated vibration data from an extrusion die assembly using DFT and MESA methods. A simulation study was performed in which the data $x_k$ were generated by passing pseudo-random binary noise $w_k$ through resonant filters with general equation of the form:

$$x_k = 2\alpha \cos(2\pi f_0)x_{k-1} - \alpha^2 x_{k-1}$$
$$+ (1-\alpha)(1+\alpha)w_k - \alpha(1-\alpha)(1+\alpha)\cos(2\pi f_0)w_{k-1} \qquad ... (6.81)$$

   where $\alpha$ was chosen to be 0.99 and one of the natural frequencies $f_0$ was chosen to be 0.25 Hz. Substitute these values into equation (6.81) and determine the filter parameters. Derive an expression for the autocorrelation function $R_{xx}(m)$, equation (6.12).

2) Check the analytical result obtained in Exercise 1 by writing a computer program to generate the $\{w_t\}$ ($\sigma_w^2 = 1$) and $\{x_t\}$ series, and compute the autocorrelation estimates $\hat{R}_{xx}(m)$.

3) Determine the power spectral density $S_{xx}(f)$ for equation (6.81) and check the analytical result by writing a computer simulation program to obtain the estimates $\hat{S}_{xx}(f)$ from the generated $\{w_t\}$ and $\{x_t\}$ series as outlined in Exercise 2.

4) Use the MESA#SIM program to perform the following simulation tests:
   a) the frequency resolution of the DFT and MESA methods as the number of 'Points/Block' is varied from 10 to 110 in steps of 20, and with 'Total Points' = 'Points/Block' (single pass);
   b) the sensitivity of the MESA estimates to AR model order by varying the 'Model Order' data value from 2 to 10 in steps of 1;

*c)* the sensitivity of the DFT and MESA methods to background noise by varying the 'Noise/Signal' data value from its default value of 0.1 to 1.

In each case, what observations can you make about the affect of these analysis parameters on the relative accuracy of the DFT and MESA spectral estimates?

5) Use the MESA#SIM program to evaluate the performance of the five model order criteria for both the Andersen and Haykin algorithms. Normalise the results you obtain from the various criteria and plot them as a function of model order (cf. Figure 6.4). What conclusions can you make about the relative merits of each criterion for determining the 'optimum' PEF order?

6) Use the MESA#SIM program to evaluate the performance of Akaike's AIC model order criterion for values of the 'Akaike Factor' ranging from 2 to 20 in increments of 2. Normalise the results you obtain from the various criteria and plot them as a function of model order (cf. Figure 6.4). What conclusions can you make about the merits of the AIC criterion for determining the 'optimum' PEF order?

# APPENDIX A6

## THE LEVINSON RECURSION

Recall from equation (6.31) that the $M$th order prediction error filter equation is given by:

$$\begin{bmatrix} R_{xx}(0) & R_{xx}(-1) & \cdots & R_{xx}(-M) \\ R_{xx}(1) & R_{xx}(0) & \cdots & R_{xx}(1-M) \\ \vdots & \vdots & \ddots & \vdots \\ R_{xx}(M) & R_{xx}(M-1) & \cdots & R_{xx}(0) \end{bmatrix} \begin{bmatrix} 1 \\ \alpha_{2:M} \\ \vdots \\ \alpha_{M:M} \end{bmatrix} = \begin{bmatrix} P_M \\ 0 \\ \vdots \\ 0 \end{bmatrix} \quad \dots \text{(A6.1)}$$

where the added subscript $M$ denotes the order, and the autocorrelation matrix $\mathbf{R}_{xx}$ is Toeplitz in form (that is, its elements $R(i,j) = R_{xx}(i-j)$; $i,j = 0, M$) and is also Hermitian ($R(i,j) = R^*(j,i)$). The Levinson recursion[1], as utilised by the Yule-Walker solution of (A6.1), exploits the Toeplitz property to proceed recursively. The following analysis is based on that of Haykin (1979).

Suppose we know the solution to the set of $M$ equations for the prediction error filter of order $M-1$, which, from equation (A6.1), is given by:

$$\sum_{k=0}^{M-1} \alpha_{k:M-1} R_{xx}(m-k) = \begin{cases} P_{M-1}, & m = 0 \\ 0 & m = 1,\dots,M \end{cases} \quad \dots \text{(A6.2)}$$

where the parameter subscript $M-1$ has been added to denote the filter order. Equation (A6.2) can be written in matrix form as:

$$\begin{bmatrix} R_{xx}(0) & R_{xx}(-1) & \cdots & R_{xx}(1-M) \\ R_{xx}(1) & R_{xx}(0) & \cdots & R_{xx}(2-M) \\ \vdots & \vdots & \ddots & \vdots \\ R_{xx}(M-1) & R_{xx}(M-2) & \cdots & R_{xx}(0) \end{bmatrix} \begin{bmatrix} 1 \\ \alpha_{2:M-1} \\ \vdots \\ \alpha_{M-1:M-1} \end{bmatrix} = \begin{bmatrix} P_{M-1} \\ 0 \\ \vdots \\ 0 \end{bmatrix} \quad \dots \text{(A6.3)}$$

Equations (A6.2) and (A6.3) represent the filter operation in the *forward* direction. We now modify equation (A6.2) to compute the filter coefficients in the *backward* direction by taking the complex conjugate of both sides and substituting $M-1-k$ for $k$ and $m$ for $M-1-m$ and obtain:

$$\sum_{k=0}^{M-1} \alpha^*_{M-1-k:M-1} R_{xx}(m-k) = \begin{cases} 0, & m = 0,1,\dots,M-2 \\ P_{M-1} & m = M-1 \end{cases} \quad \dots \text{(A6.4)}$$

---

[1]  N. Levinson's original recursion was published in 1947, and was later modified by J. Durbin in 1959. The resulting recursion is termed the Levinson-Durbin or Durbin-Levinson recursion in some texts (see Proakis *et al.* 1992).

which may be written in matrix form as:

$$
\begin{bmatrix}
R_{xx}(0) & R_{xx}(-1) & \cdots & R_{xx}(1-M) \\
R_{xx}(1) & R_{xx}(0) & \cdots & R_{xx}(2-M) \\
\vdots & \vdots & \ddots & \vdots \\
R_{xx}(M-1) & R_{xx}(M-2) & \cdots & R_{xx}(0)
\end{bmatrix}
\begin{bmatrix}
\alpha_{M-1:M-1} \\
\alpha_{M-2:M-1} \\
\vdots \\
1
\end{bmatrix}
=
\begin{bmatrix}
0 \\
0 \\
\vdots \\
P_{M-1}
\end{bmatrix}
\quad \ldots (A6.5)
$$

We combine equations (A6.3) and (A6.5) to augment the number of prediction error equations by one, as follows:

$$
\begin{bmatrix}
R_{xx}(0) & R_{xx}(-1) & \cdots & R_{xx}(1-M) & R_{xx}(-M) \\
R_{xx}(1) & R_{xx}(0) & \cdots & R_{xx}(2-M) & R_{xx}(1-M) \\
\vdots & \vdots & \ddots & \vdots & \vdots \\
R_{xx}(M-1) & R_{xx}(M-2) & \cdots & R_{xx}(0) & R_{xx}(-1) \\
R_{xx}(M) & R_{xx}(M-1) & \cdots & R_{xx}(1) & R_{xx}(0)
\end{bmatrix}
$$

$$
\times \left\{
\begin{bmatrix}
1 \\
\alpha_{1:M-1} \\
\vdots \\
\alpha_{M-1:M-1} \\
0
\end{bmatrix}
+ \rho_M
\begin{bmatrix}
0 \\
\alpha_{M-1:M-1} \\
\vdots \\
\alpha_{1:M-1} \\
1
\end{bmatrix}
\right\}
=
\left\{
\begin{bmatrix}
P_{M-1} \\
0 \\
\vdots \\
0 \\
\Delta_M
\end{bmatrix}
+ \rho_M
\begin{bmatrix}
\Delta_M^* \\
0 \\
\vdots \\
0 \\
P_{M-1}
\end{bmatrix}
\right\}
\quad \ldots (A6.6)
$$

where the quantities $\rho_M$ are sometimes called the *partial correlation coefficients*, and their negative values $-\rho_M$ the *reflection coefficients*. The partial correlation $\rho_m$ is defined as the correlation between $x_m$ and $x_{m+n}$ when $x_m, \ldots, x_{m+n-1}$ are held fixed. The AR filter is stable if the roots of $A_p(z)$ are *inside* the unit circle; that is, if $|\rho_m| < 1$. This condition is satisfied when the autocorrelation matrix, $\mathbf{R}_{xx}$, is positive definite, as noted in section 6.3.5.

Comparing equations (A6.6) with (A6.1) for the corrreponding $M$th order prediction error filter, we obtain Levinson's recursion:

$$
\alpha_{m:M} = \alpha_{m:M-1} + \rho_M \alpha_{M-m:M-1}, \quad m = 0,1,\ldots,M \quad \ldots (A6.7)
$$

together with

$$
P_M = P_{M-1} + \rho_M \Delta_M^* \quad \ldots (A6.8)
$$

and

$$
0 = \Delta_M + \rho_M P_{M-1} \quad \ldots (A6.9)
$$

We can eliminate $\Delta_M$ from equations (A6.8) and (A6.9) by taking the complex conjugate of equation (A6.9) and substituting for $\Delta_M^*$ in equation (A6.8) to obtain the more usual form:

$$P_M = P_{M-1}\left[1 - |\rho_M|^2\right] \qquad \qquad \text{... (A6.10)}$$

Note that, in Levinson's recursion formula, equation (A6.7) is

$$\alpha_{m:M} = \begin{cases} 1, & \text{for } m = 0 \\ \rho_M, & \text{for } m = M \\ 0, & \text{for } m > M \end{cases} \qquad \text{... (A6.11)}$$

**Initialisation**:

$$M = 0 \rightarrow P_0 = R_{xx}(0); \; \alpha_{0:0} = 1$$

$$M = 1 \rightarrow P_1 = P_0\left[1 - |\rho_1|^2\right]; \; \alpha_{0:1} = 1; \; \alpha_{1:1} = \rho_1 = -\frac{R_{xx}(1)}{R_{xx}(0)} \qquad \text{... (A6.12)}$$

from which successive recursions follow.

# APPENDIX B6

## MESA#SIM PROGRAM INSTALLATION AND OPERATION

MESA#SIM.EXE computes the parametric spectral estimates for one or two sinusoids in the presence of noise. It has been designed to run under Windows 3.1 with video display settings of 640×480 pixels×256 colours, and also runs under Windows 95 with the same video display settings. Refer to the **readme.txt** document for further information.

## B6.1 INSTALLATION

Perform the following steps if the executable files supplied with this book have not been previously installed.

1) Insert the floppy disc in the FDD and type **a:\setup filename** in the Windows 3.1 Program Manager File/RUN menu.
2) The **setup.bat** file copies *all* the files on the floppy disc to the hard drive directory **c:\filename**. (NOTE: The setup procedure will terminate if **filename** is not specified.)
3) The **setup.bat** file also decompresses the RUN library files **cable.dll** and **rlzrun20.rts** and instals them into a newly created Windows sub-directory **c:\windows\rlzrun20**.
4) For ease of operation, it is recommended that the four simulation programs in the **c:\filename** directory be installed as a program group icon with their separate program icons using the Windows Program Manager.

## B6.2 OPERATION

1) Double click on the icon to run the program. A Title Form is displayed and a Run option is added to the Menu Bar.
2) Program operation is straight forward and the data input options are discussed below. Choose the Data File **Default** button (for default data) then the **OK** button or **Enter** key to continue operation.
3) There is a short waiting period as the program generates a random series of NTOTAL data points. The first block of NDATA points for the times series $\{x_t\}$ are displayed (blue line) in the top chart, then the corresponding prediction error filter estimates $\{\hat{x}_t\}$ are displayed (red line). The program then computes the spectral density estimates by standard DFT methods and by Haykin's or Andersen's MESA algorithm, and displays these in separate charts below the time series chart.

4) Ctrl+C invokes a dialogue box which enables the user to terminate execution and return to the group window.
5) Use the Run menu options to execute a New Case or Exit from the program normally. The File Menu/Exit option also exits from the program and returns control to the group window.

**B6.3     SIMULATION AND OPERATIONAL OPTIONS**

The program's simulation and operational options include the following:

1) Generation of one (one gain zero) or two sine waves contaminated by zero mean random noise with variable signal to noise ratios (SNRs).
2) NDATA points/block of data are used to calculate the spectral density estimates, and the computation steps along NBLOCK points until NTOTAL data points have been analysed as disjoint time series. For model order optimisation (OPTIMISE>0), NTOTAL is set equal to NDATA and NBLOCK is set to zero by default.
3) Selection of one of the five model order criteria discussed in section 6.4 by setting the Order Criterion data value from 1→5 as follows:
$$1 = \text{COD}; 2 = \text{FPE}; 3 = \text{AIC}; 4 = \text{BIC}; 5 = \text{MDL}$$
An Akaike Factor enables greater weighting to be given to the model complexity, as discussed in section 6.4.4 (Note: the default value is 2 for the AIC and 16 for the BIC.)
4) The plotted data are printed to a Print Log which can be viewed by accessing the Window/Show Print Log (Ctrl+P) menu option. The data can be copied to the Clipboard using the Edit/Copy Window menu option or by highlighting the printed text and pressing Ctrl+Ins. The copied data can then be inserted as text in a word processor document or in a speadsheet (such as MS Excel). Alternatively, the plotted data can be printed directly using the File/Print (Ctrl+Shift+F12) menu options. The printed data includes the following arrays:

X(t)     the generated time series $\{x_t\}$;

A(t)     the estimated time series $\{\hat{x}_t\}$;

FREQ   the discrete frequency values ($f_k$; $k = 0, \text{NFREQ}$);

DFT     the DFT spectral density estimates ($S_{XX}(f_k)$; $k = 0, \text{NFREQ}$);

MESA   the MESA spectral density estimates;

AR(m)   the AR model parameters ($\alpha_m$; $m = 1, M$)

MOC     the selected model order criterion values ($m = 1, M$).

This information can also be stored on disc as a text file with a filename specified by the user as data input. However, if a null string is inserted in the data text box, the data are not stored on file.

5) The input data form is minimised when the OK button is pressed, and may be restored or maximised for viewing or printing (File/Print). The input data can be saved by pressing the SAVE AS button, and selecting an existing filename or entering a newly created filename from the menu list box.

## B6.4 FLOW CHART OF HAYKIN MESA ALGORITHM

**Data Input:**
$$\{x_t\}; t = 0,\ldots,N$$

**Initialisation:**
$$M = 0$$
$$P_0 = R_{xx}(0) = 0$$
$$P_0 = P_0 + |x_t|^2 \; ; \; t = 0,\ldots,N$$
$$P_0 = P_0 / (N+1) \; ; \; R_{xx}(0) = P_0$$
$$M = 1$$

**Algorithm:**
$$A = D = 0$$
$$A = A + e_{t-1:M-1}^{f} e_{t-1:M-1}^{b*} \; ; \qquad t = M+1,\ldots,N$$
$$D = D + \left|e_{t-1:M-1}^{f}\right|^2 + \left|e_{t-1:M-1}^{b}\right|^2 \; ; \; t = M+1,\ldots,N$$
$$\rho_M = \alpha_{M:M} = -2A/D$$
$$P_M = P_{M-1}\left(1 - |\rho_M|^2\right)$$
$$R_{xx}(M) = -\alpha_{M-m:M} R_{xx}(m) \; ; \; m = 1,\ldots,M-1$$
$$\text{FPE}(M) = P_M(N + M + 1)/(N - M - 1)$$

**IF $M = 1$ THEN**
$$TH = \text{FPE}(M)$$
$$M = M + 1$$
$$e_{t:M}^{f} = e_{t:M-1}^{f} + \rho_M e_{t-1:M-1}^{f} \; ; \; t = M+1,\ldots,N$$
$$e_{t:M}^{b} = e_{t:M-1}^{b} + \rho_M e_{t-1:M-1}^{b} \; ; \; t = M+1,\ldots,N$$

**ELSE**
$$\alpha_{m:M} = \alpha_{m:M-1} + \rho_M \alpha_{M-m:M-1}^{*} \; ; \; m = 1,\ldots,M-1$$

**IF $\text{FPE}(M) \leq TH$ THEN**

**ELSE $\rightarrow$ Output:**
$$M = M_{opt}$$
$$\alpha_{m:M} \; ; \; m = 1,\ldots,M$$
$$P_M \; ; \; R_{xx}(M)$$

# 7

# MULTIVARIATE PROCESS ANALYSIS

## 7.1    INTRODUCTION AND LEARNING OBJECTIVES

The SISO process model discussed in Chapter 4 is a fundamental component (or building block) of the more general Multiple-Input-Multiple-Output (MIMO) process model. Although the various transfer function relationships in MIMO processes may be estimated from the set of linear equations containing all the relevant spectral estimates, they cannot be estimated in the minimum least squares sense unless certain constraints are placed on the matrix of residual noise spectra. These constraints produce a set of normal equations in which the MIMO model reduces to a set of independent Multiple-Input-Single-Output (MISO) subprocesses. This chapter discusses MIMO processes and their reduction to MISO subprocesses.

The procedures that we have discussed for identifying SISO processes provide a basis for identifying higher order processes with multiple inputs and multiple outputs. Specifically, the *learning objectives* of this Chapter are to:

- **show** how general, linear MIMO processes with $M$ input signals and $Q$ output signals are analysed using *multivariate spectral analysis* methods;
- **derive** the 'optimum' input-output transfer function estimates from the set of '*normal*' MIMO process equations obtained by the imposition of *minimum least squares* and *orthogonality* conditions;
- **show** how the resulting set of normal MIMO process equations reduces the analysis problem to an equivalent set of $Q$ independent MISO subprocesses;
- **prove** that the normal MIMO process equations of the MIMO process have the property that their covariance (or spectral) matrix of residuals is diagonal;
- **define** higher order *partial spectral density functions* needed to compute the optimum input-output *transfer function estimates, partial* and *multiple coherence estimates*;
- **derive** a computationally efficient *algorithm* for obtaining the optimum input-output transfer function estimates from their respective partial spectral density functions;

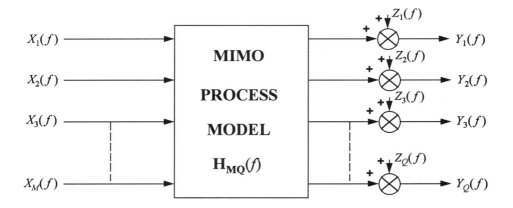

**Figure 7.1** Block diagram of MIMO process model

- **provide** some *essential guidelines* for obtaining realiable estimates of the partial spectral density functions.

These learning objectives will be supported by worked examples and simulation exercises using the software provided with this book.

## 7.2 MULTIPLE-INPUT-MULTIPLE-OUTPUT (MIMO) PROCESSES

### 7.2.1 Generalised Process Equations

A typical MIMO process with $M$ inputs, $X_i(f)$, $i = 1, M$ and $Q$ outputs, $Y_q(f)$, $q = 1, Q$ is depicted by the block diagram shown in Figure 7.1, where $Z_q(f)$ denotes the residual noise spectra superimposed on the outputs. For this generalised MIMO process, the $k$th output is *linearly* related to the $M$ inputs by the equation:

$$Y_q(f) = \sum_{j=1}^{M} X_j(f) H_{jq}(f) + Z_q(f) \qquad \dots (7.1)$$

which may be written in matrix notation as:

$$\mathbf{Y_Q'}(f) = \mathbf{X_M'}(f) \mathbf{H_{MQ}}(f) + \mathbf{Z_Q'}(f) \qquad \dots (7.2)$$

where $\mathbf{Y_Q'}(f)$ is a $1 \times Q$ row vector of outputs, $\mathbf{X_M'}(f)$ is a $1 \times M$ row vector of inputs, $\mathbf{H}_{MQ}(f)$ is the $M \times Q$ matrix of input-input and input-output transfer functions, and $\mathbf{Z}_Q(f)$ is the $1 \times Q$ row vector of uncorrelated noise sources.

Following the procedures outlined in Chapter 4, premultiplying equation (7.2) by the complex conjugate column vector of inputs $\mathbf{X}_M^*(f)$ and taking expectations yields the matrix spectral equation:

$$\mathbf{S}_{MQ}(f) = \mathbf{S}_{MM}(f)\mathbf{H}_{MQ}(f) + \Xi_{MQ}(f) \qquad \text{... (7.3)}$$

where $\Xi_{MQ}(f)$ is an $(M \times Q)$ cross-spectral matrix relating the inputs and residual noise spectra on the outputs (cf. equation (4.13), Chapter 4). It follows from equation (4.14) that, ideally, $\Xi_{MQ}(f) = 0$, and hence the transfer function relationships between the various inputs and outputs are given by the matrix equation (cf. equation (4.11)):

$$\mathbf{H}_{MQ}(f) = \mathbf{S}_{MM}^{-1}(f)\mathbf{S}_{MQ}(f) \qquad \text{... (7.4)}$$

Equation (7.4) is computationally intensive and inefficient, and may not yield optimum transfer function estimates in the minimum least squares sense, depending on the validity of the assumption that $\Xi_{MQ}(f) \to 0$ for all frequencies. A more efficient analysis method in which the MIMO process is reduced to $Q$ independent MISO subprocesses, is outlined in the following section.

### 7.2.2       Reduction of MIMO Processes to MISO Processes

The $M$th-order residual $Z_{p:M}(f)$ on the $p$th MIMO process output $Y_p(f)$, $p=1,Q$, is linearly related to the inputs, $X_i(f)$ $(i=1,M)$, by:

$$Z_{p:M}(f) = Y_p(f) - \sum_{i=1}^{M} X_i(f)H_{ip}(f) \qquad \text{... (7.5)}$$

where $H_{ip}(f)$ is the transfer function between the $i$th input and the $p$th output.

Similarly, the $M$th-order residual on the $q$th process output, $Y_q(f)$ is:

$$Z_{q:M}(f) = Y_q(f) - \sum_{k=1}^{M} X_k(f)H_{kq}(f) \qquad \text{... (7.6)}$$

The partial cross-spectrum between the $p$th and $q$th residual spectra is defined by (assuming frequency dependence of the variables):

$$P_{pq:M}(f) = \lim_{T \to \infty} \frac{1}{T}[Z^{\cdot}_{p:M}(f)Z_{q:M}(f)]$$

... (7.7)

$$= S_{pq} - \sum_{k}^{M} S_{pq} H_{kq} - \sum_{i}^{M} S_{iq} H^*_{ip} + \sum_{i}^{M}\sum_{k}^{M} S_{ik} H^*_{ip} H_{kq}$$

where (*) denotes the complex conjugate. Note that the covariance between the $p$th and $q$th residuals is

$$\sigma^2_{pq:M} = \int_{-\infty}^{\infty} P_{pq:M}(f)\,df$$

... (7.8)

As with SISO processes (Chapter 4, section 4.3.2), we use least squares to determine the necessary conditions for identifying the input-output transfer function estimates. In a similar manner to equation (4.14), we first divide the transfer functions, $H_{mr}(f)$ say, in equation (7.7) into its orthogonal real and imaginary components, $H_{mr}(f) = A_{mr}(f) - jB_{mr}(f)$. We then differentiate the resulting equation with respect to the orthogonal real and imaginary components of the transfer function between the $m$th input and $r$th output (cf. equation (4.9)):

$$-\sum_{k}^{M} S_{pk} \frac{\partial A_{kq}}{\partial A_{mr}} - \sum_{i}^{M} S_{iq} \frac{\partial A_{ip}}{\partial A_{mr}} + \sum_{i}^{M}\sum_{k}^{M} S_{ik}\left\{\frac{\partial A_{ip}}{\partial A_{mr}} H_{kq} + \frac{\partial A_{kq}}{\partial A_{mr}} H^*_{ip}\right\} = 0$$

... (7.9)

$$\sum_{k}^{M} S_{pk} \frac{\partial B_{kq}}{\partial B_{mr}} - \sum_{i}^{M} S_{iq} \frac{\partial B_{ip}}{\partial B_{mr}} + \sum_{i}^{M}\sum_{k}^{M} S_{ik}\left\{\frac{\partial B_{ip}}{\partial B_{mr}} H_{kq} - \frac{\partial B_{kq}}{\partial B_{mr}} H^*_{ip}\right\} = 0$$

Equations (7.9) have no unique solution unless the MIMO process equations are normal; that is, unless we impose the *orthogonality condition*:

$$\frac{\partial A_{ns}}{\partial A_{mr}} = \frac{\partial B_{ns}}{\partial B_{mr}} = 1; \quad m=n, r=s$$

... (7.10)

$$= 0; \quad\quad\quad m \neq n, r \neq s$$

Applying this condition to equations (7.9) and adding the resulting equations yields the spectral relationship between the $i$th input and the $q$th output as:

$$S_{iq}(f) = \sum_{k=1}^{M} S_{ik}(f) H_{kq}(f)$$

... (7.11)

which, for $M$ inputs, may be written in matrix notation as

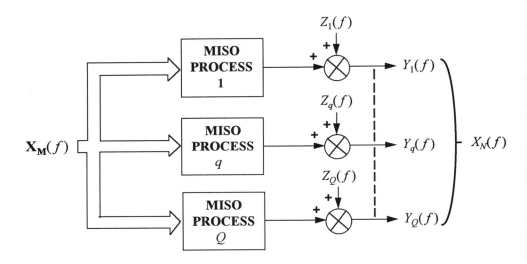

**Figure 7.2** Reduction of MIMO process to $Q$ independent MISO sub-processes

$$\mathbf{S_M}(q,f) = \mathbf{S_{MM}}(f)\mathbf{H_M}(q,f) \qquad\qquad \text{... (7.12)}$$

where $\mathbf{S_M}(q,f)$ and $\mathbf{H_M}(q,f)$ are $M$-dimensional column vectors and $\mathbf{S_{MM}}(f)$ is an ($M{\times}M$) matrix of spectral functions relating *all* the inputs. Equation (7.12), from which the minimum least squares estimates of the process transfer functions $H_{kq}(f)$ relating the $M$ inputs, $X_k(f)\,(k=1,M)$, and the $q$th output, $X_q(f)$, are evaluated, is the MISO subprocess equivalent of equation (4.16) for SISO processes.

Substitution of equation (7.11) into equation (7.7) shows that minimum least squares estimates of the input-output process transfer functions $H_{kq}(f)$ are obtained when the residual spectral (or covariance) matrix is *diagonal*; that is, when the MIMO process satisfies the condition:

$$
\begin{aligned}
P_{pq:M}(f) &= 0 && p \neq q \\
&= P_{pp:M}(f) && p = q
\end{aligned}
\qquad\qquad \text{... (7.13)}
$$

This condition means that the residual noise spectra, $Z_{q:M}(f)$ and $Z_{p:M}(f)$, superimposed on their respective outputs must be independent/uncorrelated ('white'), and provides a test for normality and the reduction of the MIMO process to $Q$ independent MISO sub-processes, Figure 7.2.

## 7.3     MULTIPLE-INPUT-SINGLE-OUTPUT (MISO) PROCESS ANALYSIS

MISO process models are perhaps the most important class of multivariate models, particularly those using spectral estimates calculated by fast Fourier transform algorithms. However, as shown in equation (7.12), traditional least squares formulae for the multivariate spectral analysis of MISO linear processes require complex matrix manipulation, and in the following analysis we develop a more efficient method for calculating the input-input and input-output transfer function estimates from their respective *partial spectral density estimates*.

### 7.3.1     Basic MISO Process Equations

A general model of a multiple input single output (MISO) linear process with $M$ inputs is schematically represented by the block diagram shown in Figure 7.3. For reasons that will become clear in the next section, we now denote any $(q$th) output of a MIMO process $(q = 1,Q)$, Figure 7.2, by a single frequency-dependent output $X_N(f)$ $(N = M + 1)$ of the MISO process, Figure 7.3. From equation (7.11), $X_N(f)$ is linearly related to the inputs $X_m(f)$ $(m = 1, M)$ in a minimum least squares sense by the equation:

$$S_{kN}(f) = \sum_{j=1}^{M} S_{jk}(f) H_{kN}(f) \qquad \qquad \text{... (7.14)}$$

We note in passing that commercial spectrum analysers are often used to analyse multivariate (MIMO/MISO) processes, and since these analysers invariably use the *ordinary* bivariate spectral analysis methods described in Chapter 4, they effectively treat multivariate processes as a SISO process by neglecting the contributions of the other $(K)$ input-output transfer relationships;

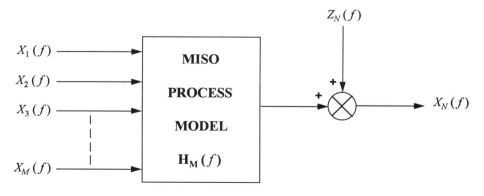

**Figure 7.3** Block diagram of MISO process model

that is, for the $k$th input-output SISO relationship

$$\underbrace{S_{kN}(f) = S_{kk}(f)H_{kN}(f)}_{\text{Bivariate (SISO) model}} + \underbrace{\sum_{\substack{j=1 \\ j \ne k}}^{M} S_{kj}(f)H_{jN}(f)}_{\text{Neglected}}, \qquad \text{... (7.15)}$$

the second term on the right hand side is neglected. This assumption is only valid for the special case where the inputs are independent; that is, $S_{kj}(f) = 0$.

Equation (7.14) may be written in matrix form as:

$$\mathbf{S_M}(f) = \mathbf{S_{MM}}(f)\mathbf{H_M}(f) \qquad \text{... (7.16)}$$

where $\mathbf{S_M}(f)$ and $\mathbf{H_M}(f)$ denote the respective ($M{\times}1$) column vectors, and $\mathbf{S_{MM}}(f)$ denotes the ($M{\times}M$) matrix of ordinary spectral density functions relating the inputs. Equation (7.16) can be formally solved to yield

$$\mathbf{H_M}(f) = \mathbf{S_{MM}^{-1}}(f)\mathbf{S_M}(f) \qquad \text{... (7.17)}$$

that is, as with the MIMO case, the solution of equation (7.17) to obtain the individual input-output transfer functions requires the inversion of the complex matrix $\mathbf{S_{MM}}(f)$. An alternative method is to compute the input-output transfer functions from their *partial spectral density functions*. This method gives more comprehensive information of the effect of each input on the output.

### 7.3.2 Partial Spectral Density Functions

The partial spectral density functions are derived from the residual (*conditioned*) noise spectra by subtracting the effects of the other $K$ ($= M{-}1$) variables from the variables of interest (Figure 7.4). Thus, we define the $K$th-order residual noise spectrum of the $k$th input, $X_k(f)$, as:

$$Z_{k:K}(f) = X_k(f) - \sum_{\substack{j=1 \\ j \ne k}}^{M} X_j(f)H_{jk}(f) \qquad \text{... (7.18)}$$

where $X_j(f)$ denotes the other inputs, $H_{jk}(f)$ denotes the transfer function between the $j$th and $k$th inputs, and the subscript of $Z_{k:K}(f)$ denotes the $k$th variable, $K$th-order residual. As we will show, the conditioned spectra $Z_{k:K}(f)$ is the residual *input* to the input-output transfer function $H_{kN}(f)$.

We define the $K$th-order conditioned spectrum, $Z_{N:K}(f)$, of the output, $X_N(f)$, in a similar manner by:

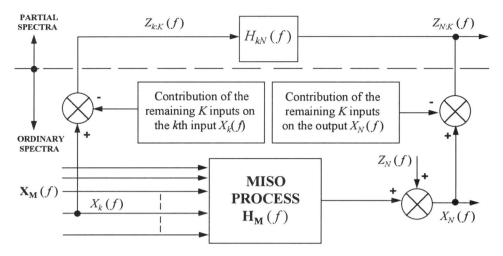

**Figure 7.4** Schematic showing computation of partial spectral density relationships

$$Z_{N:K}(f) = X_N(f) - \sum_{\substack{m=1 \\ m \neq k}}^{M} X_m(f) H_{mN}(f) \qquad \text{... (7.19)}$$

where $H_{mN}(f)$ denotes the transfer function between the $m$th input and the output ($N$), but which excludes the $k$th input-output relationship ($m \neq k$). An alternative form of (7.19) is derived from equation (7.1) as follows:

$$X_N(f) = \sum_{m=1}^{M} X_m(f) H_{mN}(f) + Z_N(f) \qquad \text{... (7.20)}$$

and hence

$$\underbrace{X_N(f) - \sum_{\substack{m=1 \\ m \neq k}}^{M} X_m(f) H_{mN}(f)}_{=Z_{N:K(f)}} = X_k(f) H_{kN}(f) + Z_N(f) \qquad \text{... (7.21)}$$

that is, $Z_{N:K}(f)$ is also given by:

$$Z_{N:K}(f) = X_k(f) H_{kN}(f) + Z_N(f) \qquad \text{... (7.22)}$$

We now compute the relevant partial spectral density functions as follows: The *input partial power spectral density function* is defined as:

$$P_{kk:K}(f) = E[Z_{k:K}^*(f) Z_{k:K}(f)] \qquad \text{... (7.23)}$$

where the superscript * denotes the complex conjugate. Substitution of equation (7.18), collecting terms and simplifying using equation (7.14) yields:

$$P_{kk:K}(f) = S_{kk}(f) - \sum_{\substack{m=1 \\ m \neq k}}^{M} S_{km}(f) H_{mk}(f) \qquad \text{... (7.24)}$$

Similarly, the *input-output partial cross-spectral density function* is defined as:

$$P_{kN:K}(f) = E[Z^*_{k:K}(f) Z_{N:K}(f)] \qquad \text{... (7.25)}$$

and substitution of equations (7.18) and (7.19) yields:

$$P_{kN:K}(f) = S_{kN}(f) - \sum_{\substack{m=1 \\ m \neq k}}^{M} S_{km}(f) H_{mN}(f) \qquad \text{... (7.26)}$$

To obtain the input-output transfer function, $H_{kN}(f)$, we multiply equation (7.22) by the complex conjugate of the $k$th residual noise function, $Z^*_{k:K}(f)$, and obtain:

$$P_{kN:K}(f) = \left[ S_{kk}(f) - \sum_{\substack{m=1 \\ m \neq k}}^{M} S_{km}(f) H_{mk}(f) \right] H_{kN}(f) \qquad \text{... (7.27)}$$

$$= P_{kk:K}(f) H_{kN}(f)$$

where we have assumed, as in the SISO case, that the $k$th input residual $Z_{k:K}(f)$ and the output residual noise $Z_N(f)$ are uncorrelated.

The $K$th-order *output partial power spectral density function* is defined as:

$$P_{NN:K}(f) = E[Z^*_{N:K}(f) Z_{N:K}(f)] \qquad \text{... (7.28)}$$

Substitution of equation (7.22) into (7.28) yields:

$$P_{NN:K}(f) = S_{NN}(f) + S_{kk}(f) |H_{kN}(f)|^2 \qquad \text{... (7.29)}$$

Finally, in defining the various *coherence functions* in the next section, we also need to know the output residual noise, $Z_N(f)$. This is given by the residual information after subtracting the contributions of *all* inputs on the output; that is,

$$Z_N(f) = Z_{N:M}(f) = X_N(f) - \sum_{m=1}^{M} X_m(f) H_{mN}(f) \qquad \text{... (7.30)}$$

The $M$th-order *output partial power spectral density function* is defined in a similar manner to equation (7.28) as:

$$P_{NN:M}(f) = E[Z^*_{N:M}(f) Z_{N:M}(f)] = E[Z^*_N(f) Z_N(f)] \qquad \text{... (7.31)}$$

The reader should note the difference in the definitions of $S_{NN}(f)$ and $P_{NN:M}(f)$.

### 7.3.3 Ordinary, Partial and Multiple Coherence Functions

The *ordinary coherence function* between the $k$th input and the output is defined by (cf. equation (4.15)):

$$\gamma^2_{kN}(f) = \frac{|S_{kN}(f)|^2}{S_{kk}(f) S_{NN}(f)} \qquad \text{... (7.32)}$$

and is a measure of the coherent information transferred between the $k$th input and the output when *the contributions of all other variables are ignored.*

The $K$th-order *partial coherence function* between the $k$th input and the output is defined as

$$\gamma^2_{kN:K}(f) = \frac{|P_{kN:K}(f)|^2}{P_{kk:K}(f) P_{NN:K}(f)} \qquad \text{... (7.33)}$$

and is a measure of the coherent information transferred between the $k$th input and the output when *allowance is made for the effects of all other inputs on both the input and the output*; that is, the coherence between the $Z_{k:K}(f)$ and $Z_{N:K}(f)$ $K$th-order residuals.

The *multiple coherence function* is defined as

$$\gamma^2(f) = 1 - \frac{P_{NN:M}(f)}{S_{NN}(f)} \qquad \text{... (7.34a)}$$

and is a *measure of the coherent information transferred between all the inputs and the output.* Alternatively, the multiple coherence function also may be written in matrix form as:

$$\gamma^2(f) = 1 - \frac{|\mathbf{S_{NN}}(f)|}{S_{NN}(f)|\mathbf{S_{MM}}(f)|} \qquad \text{... (7.34b)}$$

where $|\mathbf{S}_{\mathrm{MM}}(f)|$ and $|\mathbf{S}_{\mathrm{NN}}(f)|$ are, respectively, the determinants of the *input* spectral density matrix, $\mathbf{S}_{\mathrm{MM}}(f)$, and the *augmented process matrix*, $\mathbf{S}_{\mathrm{NN}}(f)$, defined below.

Note that we can also use equations (7.32) and (7.33) to compute the *ordinary* and *partial* coherence between the *k*th input and any other (*m*th) input by setting $N = m$ in their respective equations. This implies that we can interchange any (*m*th) input with the designated output ($N$) to investigate the merits of any sub-order MISO process models which may have some physical significance (see section 7.3.5).

### 7.3.4    Algorithm Recurrence Equation

The computation of the input-output transfer functions, $H_{kN}(f)$, from their ordinary spectral density functions, equation (7.17), or from the partial spectral density functions, equation (7.27) involves the inversion of complex matrices. In this section we develop an *algorithmic* method which obviates the need for cumbersome complex matrix arithmetic, and is more practical for on-line process identification studies.

We begin by expressing the MISO process output, $X_N(f)$, as a linear function of the *m*th residual spectra, $Z_{mN:m-1}(f)$, Figure 7.5, by the equation:

$$X_N(f) = \sum_{m=1}^{M} Z_{m:m-1}(f) F_{mN}(f) + Z_N(f) \qquad \dots (7.35)$$

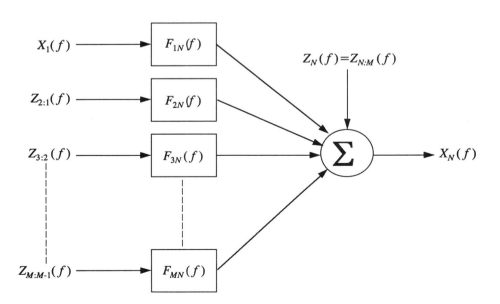

**Figure 7.5** MISO process model for residual spectra

where $F_{mN}(f)$ is the transfer function relating the $m$th and $N$th residual (conditioned) spectra, and is related to the process transfer functions by:

$$F_{mN}(f) = \sum_{j=m}^{M} F_{mj}(f) H_{jN}(f)$$

$$H_{mN}(f) = F_{mN}(f) - \sum_{j=m+1}^{M} F_{mj}(f) H_{jN}(f)$$

... (7.36)

If $m$ inputs are subtracted consecutively from the $p$th variable as shown in Figure 7.6, then the $m$th-order residual $Z_{p:m}(f)$ is

$$Z_{p:m}(f) = Z_{p:l}(f) - Z_{n:l}(f) F_{np}(f), \quad l=m-1; \ p=m+1, N \qquad \text{... (7.37)}$$

Premultiplying equation (7.37) by the complex conjugate of the $n$th input residual, $Z_{n:l}^{*}(f)$, gives, using the definitions in equations (7.24) and (7.26),

$$P_{np:l}(f) = P_{nn:l}(f) F_{np}(f) \qquad \text{... (7.38)}$$

in which $Z_{n:l}(f)$ and $Z_{p:m}(f)$ are assumed to be uncorrelated; that is, $F_{np}(f)$ is the transfer function with minimum mean square error.

Similarly, the $m$th-order residual spectrum $Z_{q:m}(f)$ for the $q$th variable is

$$Z_{q:m}(f) = Z_{q:l}(f) - Z_{n:l}(f) F_{nq}(f) \qquad \text{... (7.39)}$$

and premultiplication by $Z_{n:l}^{*}(f)$ gives

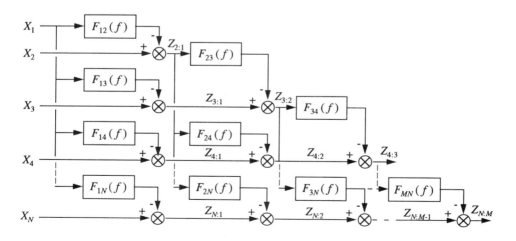

**Figure 7.6** Computation of residual spectra from measured spectra

$$P_{nq:l}(f) = P_{nn:l}(f) F_{nq}(f) \qquad \ldots (7.40)$$

The $m$th-order partial cross-spectral density function between the $p$th and $q$th variables is

$$P_{pq:m}(f) = E[Z_{p:m}^*(f) Z_{q:m}(f)] \qquad \ldots (7.41)$$

Substitution of equations (7.37) and (7.39) gives, using equations (7.38) and (7.40),

$$P_{pq:m}(f) = P_{pq:l}(f) - \frac{P_{pn:l}(f) P_{nq:l}(f)}{P_{nn:l}(f)} \qquad \ldots (7.42)$$

in which $l = m-1$, $n = p-1$, $m$, $p = 1,2,\ldots,M$ and $q = m+1,m+2,\ldots,N$. This basic *recurrence equation* relates the $m$th-order partial spectral density function to the $l$th, or $(m-1)$th, order partial spectral density functions calculated in the previous step.

From equation (7.42), the first order partial spectral density function is, with $(m = 1)$:

$$P_{pq:1}(f) = S_{pq}(f) - \frac{S_{pn}(f) S_{nq}(f)}{S_{nn}(f)} \qquad \ldots (7.43)$$

where $S_{pn}(f)$, etc., are elements of the *augmented spectral density matrix*, $\mathbf{S}_{NN}(f)$, of input-input and input-output ordinary spectral density functions:

$$\mathbf{S}_{NN}(f) = \begin{bmatrix} S_{11}(f) & \cdots & S_{1M}(f) & S_{1N}(f) \\ \vdots & \ddots & \vdots & \vdots \\ S_{M1}(f) & \cdots & S_{MM}(f) & S_{MN}(f) \\ S_{N1}(f) & \cdots & S_{NM}(f) & S_{NN}(f) \end{bmatrix} \qquad \ldots (7.44)$$

## 7.4      COMPUTATIONAL PROCEDURES

### 7.4.1      Stepwise Manipulation of the Augmented Spectral Matrix

In this section, we manipulate the above $(N \times N)$ augmented matrix, $\mathbf{S}_{NN}(f)$, by repeated application of the algorithm equation (7.42) and concatenate it to a $(2 \times 2)$ partial spectral density matrix $\mathbf{P}_{kN:22}(f)$:

$$\mathbf{P}_{kN:22}(f) = \begin{bmatrix} P_{kk:K}(f) & P_{kN:K}(f) \\ P_{Nk:K}(f) & P_{NN:K}(f) \end{bmatrix} \qquad \ldots (7.45)$$

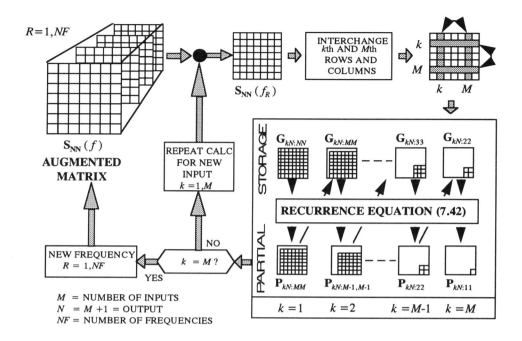

**Figure 7.7** Schematic diagram of computational procedure

relating the $K$th-order input-input, input-output and output partial spectral density functions between the $k$th input and the output ($N$), equations (7.24) to (7.27).

A schematic diagram of the computational procedure is depicted in Figure 7.7, and it consists of the following steps.

1) For a specified frequency $f_R$, an ($N \times N$) storage matrix $\mathbf{G}_{kN:NN}$ is set up in which the $k$th and $M$th rows and columns of the augmented matrix $\mathbf{S}_{NN}(f_R)$ are interchanged.

2) The first order partial spectral density functions are calculated from the elements of the storage matrix $\mathbf{G}_{kN:NN}$ using the recurrence equation (7.42), and are stored in the ($M \times M$) partial matrix $\mathbf{P}_{kN:MM}(f_R)$.

3) ($M \times M$) elements of the storage matrix $\mathbf{G}_{kN:NN}$ are replaced by the elements of the ($M \times M$) partial matrix $\mathbf{P}_{kN:MM}(f_R)$, and become the partial spectral density functions on the right hand side of equation (7.42).

4) The second and higher order partial spectral density functions are obtained by repeating steps (2) and (3) for $K = 2,3,...,M$ as shown in Figure 7.7.

5) When $K = M-1$, the elements $P_{kN:K}(f_R)$, $P_{kk:K}(f_R)$, and $P_{NN:K}(f_R)$ of the $(2 \times 2)$ partial spectral density matrix $\mathbf{P_{kN:22}}(f_R)$ are obtained, and these are used to calculate the transfer function $H_{kN}(f_R)$ and the corresponding partial coherence function $\gamma^2_{kN:K}(f_R)$, equations (7.27) and (7.33) respectively.

6) The final step in the calculation ($K = M$) yields the power spectral density function $P_{NN:M}(f_R)$ of the residual noise $Z_N(f_R)$ on the output, and is used to calculate the multiple coherence function $\gamma^2(f_R)$, equation (7.34).

7) Steps 1 to 6 are repeated for each frequency to give the complete spectrum of results.

The above computational procedure has the advantage that it is much more efficient in computer storage and time than traditional methods, which require more (complex) matrix storage and manipulation. Another advantage of the present method is its ability to give valuable insight into the infrastructure of the MISO process, particularly with regard to the degree of influence the inputs have on each other and on the output. This desirable flexibility is demonstrated by noting that:

a) the effect of a variable on the $k$th variable and the output can be evaluated at each step in the calculation by inspection of the $(2 \times 2)$ partial sub-matrix, $\mathbf{P_{kN:22}}(f_R, K)$, for each value of $K$;

b) the relationship between the $j$th and $k$th inputs can be obtained by interchanging the $j$th and $(M-1)$th rows and columns, and for $K = M-2$ the resultant $(3 \times 3)$ partial matrix is

$$\mathbf{P_{kN:33}}(f) = \begin{bmatrix} P_{jj:K}(f_R) & P_{jk:K}(f_R) & P_{jN:K}(f_R) \\ P_{kj:K}(f_R) & P_{kk:K}(f_R) & P_{kN:K}(f_R) \\ P_{Nj:K}(f_R) & P_{Nk:K}(f_R) & P_{NN:K}(f_R) \end{bmatrix} \qquad \dots (7.46)$$

from which the transfer function $H_{jk}(f_R)$ and partial coherence function $\gamma^2_{jk:K}(f_R)$ are calculated from the partial spectral density functions $P_{jj:K}(f_R)$, $P_{jk:K}(f_R)$ and $P_{kk:K}(f_R)$ as previously; and alternatively

c) the $k$th input can be considered as the output of a MISO model in which the other $(M-1)$ variables are the inputs, the $j$th and $(M-1)$th rows and columns are interchanged as previously, and the resultant $(2 \times 2)$ partial matrix $\mathbf{P}_{jk:22}(f_R)$ is used to calculate the transfer and coherence functions $H_{jk}(f_R)$ and $\gamma^2_{jk:K}(f_R)$ respectively.

### 7.4.2    Some Practical Considerations

The following practical considerations should be noted in computing the higher order partial spectral density functions using the above algorithm and computational procedures.

1) *Coherence Between Inputs*: As a 'rule-of-thumb', the *ordinary* coherence between any two inputs should not be greater than 0.8 at any given frequency, otherwise the computed *partial* coherence estimates may become spuriously high ($\gg 1$). This is an indication that the assumed multivariate process model is approaching a *singularity* at that frequency, and is resolved by eliminating one of the measured inputs from the model. Ideally, for the most reliable input-output partial spectral density and hence MISO transfer function estimates, the inputs should be *independent*, as shown in Figure 7.8. This means that, mathematically, the spectral matrix of inputs, $\mathbf{S}_{MM}(f)$, should ideally be *diagonal* at all frequencies for best results. As we have noted, however, and as shown in equation (7.15), this is a special case of more general MISO process analysis in which the input-output transfer functions can be computed from their corresponding *ordinary* spectral density functions. The differences between the transfer function estimates obtained by the two analyses will be highlighted by discrepancies between their respective *ordinary* and *partial* coherence functions, $\gamma^2_{kN}(f)$ and $\gamma^2_{kN:K}(f)$, equations (7.32) and (7.33).

2) *Data Filtering*: As shown in section 4.2.6 of Chapter 4, *identical* filters should be used to filter *all* measurements. Filters with dissimilar frequency response characteristics will produce phase distortions in the *ordinary* (SISO process) spectral density estimates, and these distortions are usually compounded in the complex vector relationships of the dependent *partial* spectral density estimates of higher order multivariate processes. The partial spectral density estimates should be computed only between the high-pass/low-pass cutoff frequency settings of the filters. These distortions in the higher order vector relationships may also result in partial and multiple coherence estimates greater than unity ($>1$).

3) *Data Smoothing*: For similar reasons, identical *window closing* and *window carpentry* (i.e. window functions) should be used in order to minimise any leakage and possible phase distortions in the higher order partial spectral density estimates (cf. section 4.2.7, Chapter 4).

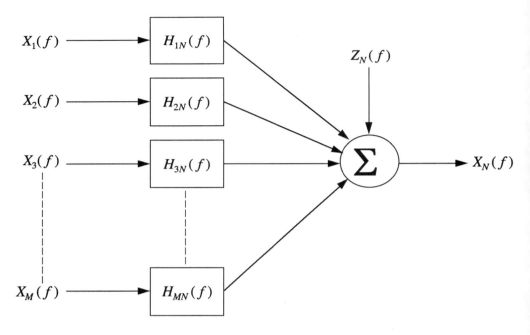

**Figure 7.8** MISO process model with independent inputs

4) ***Time Delays***: Likewise, for similar reasons, multivariate processes are more sensitive to time delays than bivariate processes, and some preprocessing is needed to determine the presence of any time delays between pairs of input-input and input-output measurements. Time delays between pairs of measurements should be compensated for using the *alignment* procedures discussed in section 4.2.8, Chapter 4.

## 7.5      A SIMULATION EXAMPLE

The following simulation example demonstrates some of the more important aspects of multivariate process analysis discussed above. The MS Windows based simulation program (MISO#SIM.EXE) is found on the demonstration disc included with this book. Some installation and operational notes on the simulation program are given in Appendix A. The structure of the MISO process model are discussed in the following section.

### 7.5.1      MISO Simulation Process

A block diagram of the MISO simulation process is depicted in Figure 7.9, where the various input-input and input-output transfer functions, $H_{jk}(s)$, the variables, $X_j(s)$, and the residual noise, $Z_4(s)$, are expressed in terms of the

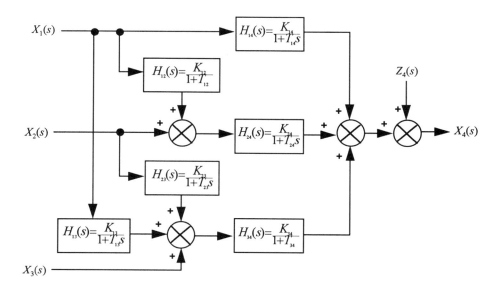

**Figure 7.9** Block diagram of MISO simulation process

Laplace operator $(s)$. The transfer functions $H_{jk}(s)$ are assumed to be first order continuous processes with static gains $K_{jk}$ and time constants $T_{jk}$, which are specified as data input. The program calculates the corresponding discrete process parameters and uses these to compute the *theoretical* discrete process transfer functions $H_{jk}(z)$. The inputs and the residual noise are zero mean, independent random sequences with amplitude values specified as data input.

The default data input values are given in Table 7.1. The input-output transfer function and coherence estimates are presented for the following cases by way of an example:

**Table 7.1** Default data input for MISO demonstration program

| | | | | | |
|---|---|---|---|---|---|
| Data Points/Step | = 128 | $K_{12}$ | = 0 | $T_{12}$ | = 0.6 |
| Total Data Points | = 2048 | $K_{13}$ | = 0 | $T_{13}$ | = 0.8 |
| Time Step | = 0.1 | $K_{14}$ | = 10 | $T_{14}$ | = 1.0 |
| Input 1 Gain | = 1 | $K_{23}$ | = 0 | $T_{23}$ | = 1.2 |
| Input 2 Gain | = 1 | $K_{24}$ | = 10 | $T_{24}$ | = 1.6 |
| Input 3 Gain | = 1 | $K_{34}$ | = 10 | $T_{34}$ | = 2.0 |
| Output Gain | = 1 | | | | |

1) *independent* inputs: the static gains of the input-input transfer functions, $K_{12}$, $K_{13}$ and $K_{23}$, are set to zero, and
2) *partially correlated* inputs: the static gains of the input-input transfer functions, $K_{12}$, $K_{13}$ and $K_{23}$, are set to unity.

## 7.5.2    MISO Simulation Results

In each case, the combinations of *ordinary* (bivariate) power and cross-spectra are computed from their respective filtered and (Hanning) windowed block time series data. The *ordinary* input-output transfer function and coherence estimates at each frequency are computed, updated and displayed each time step. The corresponding *partial* input-output transfer function and coherence estimates, as well as the *multiple* coherence estimates, are computed and displayed when the number of block time steps (= Total points/Block points) are completed.

### 7.5.2.1    Independent Inputs

This case is the one *generally assumed* in practice. Select the **Default** data button (input-input static gains $K_{12} = K_{13} = K_{23} = 0$). On completion of execution, note from the plots that (see Figures 7.10):

a) the *partial* transfer function magnitude estimates compare favourably with their corresponding *theoretical* transfer functions, whereas those computed from the *ordinary* spectral density estimates exhibit greater deviation, particularly in the vicinity of the zero frequency;
b) the *partial* transfer function phase estimates also compare favourably with their corresponding *theoretical* values, whereas those computed from the *ordinary* spectral density estimates exhibit greated deviation, particularly at the higher frequencies;
c) the computed *partial* coherence estimates, however, are considerably higher than the *ordinary* coherence estimates, thus resulting in tighter *confidence limits* associated with the more accurate *partial* transfer function magnitude and phase estimates.

The last point is of paramount importance when assessing the merits of multivariate MISO process models over ordinary SISO process models, and are a useful guide for determining which inputs are *redundant* (on a 'black box' basis) to the analysis of the multivariate process. Recall from section 4.5, Chapter 4, that the $100(1-\alpha)\%$ confidence intervals for the *ordinary* transfer function modulus and phase estimates between the $i$th and $j$th variables are respectively:

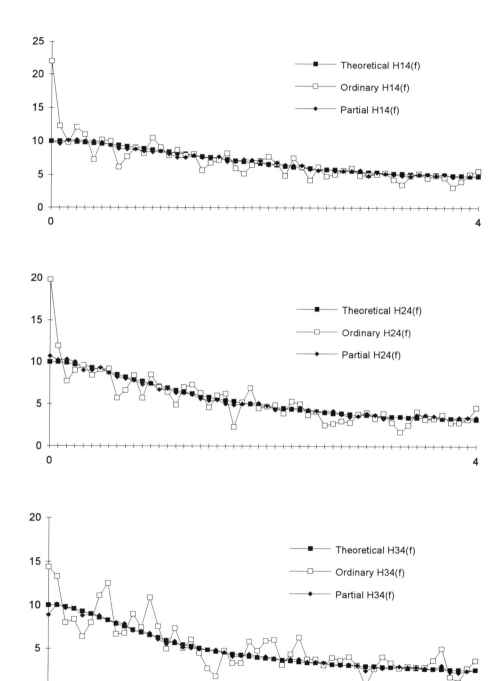

**Figure 7.10***a*  Transfer function modulus estimates – independent inputs

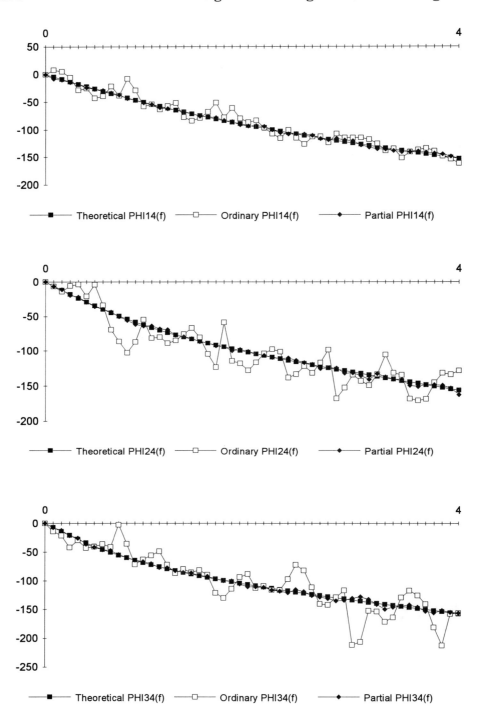

**Figure 7.10***b*  Transfer function phase estimates (deg) – independent inputs

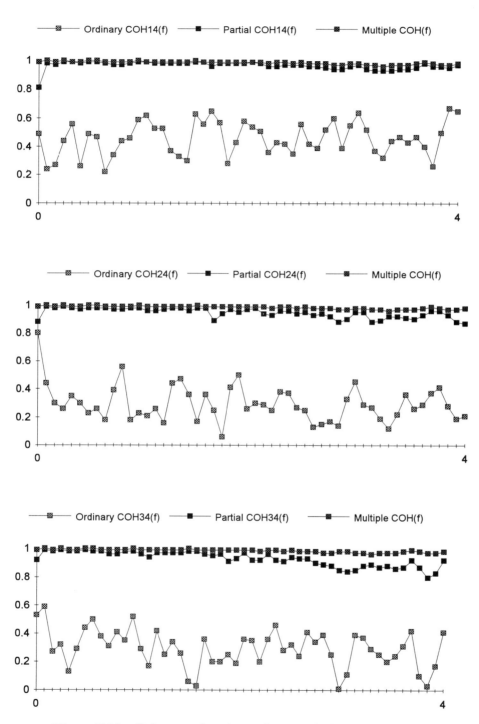

**Figure 7.10***c*  Coherence function estimates – independent inputs

$$\left|\hat{H}_{ij}(f)\right|\{1\pm r_{ij}(f)\}$$
$$\hat{\phi}_{ij}(f)\pm\sin^{-1}\{r_{ij}(f)\} \qquad \qquad \dots (7.47)$$

where the radius of the 'confidence' circle, $r_{ij}(f)$, about the estimate $\hat{H}_{ij}(f)$ is given by:

$$r_{ij}(f)=\sqrt{\frac{2}{\eta-2}f_{2,\eta-2}(1-\alpha)\frac{1-\gamma_{ij}^2(f)}{\gamma_{ij}^2(f)}} \qquad \qquad \dots (7.48)$$

The confidence intervals for the *partial* transfer function modulus and phase estimates are defined in a similar manner by equation (7.47), except that the corresponding *partial* confidence circle radius, $r_{kN:K}(f)$, is given by:

$$r_{kN:K}(f)=\sqrt{\frac{2}{\eta-2}f_{2,\eta-2}(1-\alpha)\frac{1-\gamma_{kN:K}^2(f)}{\gamma_{kN:K}^2(f)}} \qquad \qquad \dots (7.49)$$

where $\gamma_{kN:K}^2(f)$, equation (7.33), is the *partial* coherence function between the $k$th input and the output ($N$).

### 7.5.2.2    Partially Correlated Inputs

This more *general case* is the one that occurs most often in practice. Create a **miso#sim.dat** data file with input-input gains $K_{12} = K_{13} = K_{23} = 1$. On completion of execution, note from the plots that (see Figures 7.11):

a) the *ordinary* transfer function modulus estimates are distorted from their corresponding true values, particularly at low frequencies, whereas the *partial* transfer function modulus estimates again compare favourably with their corresponding *theoretical* transfer functions;

b) the *ordinary* transfer function phase estimates deviate markedly from their corresponding true values, particularly at the higher frequencies, whereas the *partial* transfer function phase estimates are in good agreement with their corresponding *theoretical* transfer functions;

c) the computed *partial* coherence estimates are again higher than the *ordinary* coherence estimates, particularly at the upper frequencies, and result in tighter *confidence limits* and, therefore, more accurate *partial* transfer function estimates.

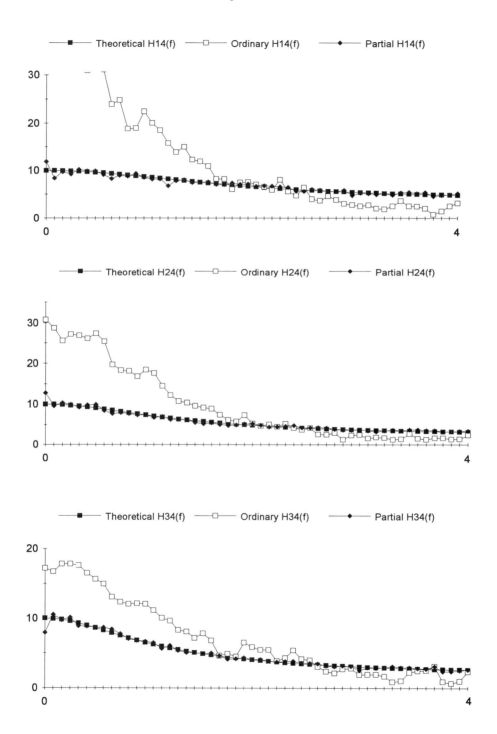

**Figure 7.11**$a$  Transfer function modulus estimates – correlated inputs

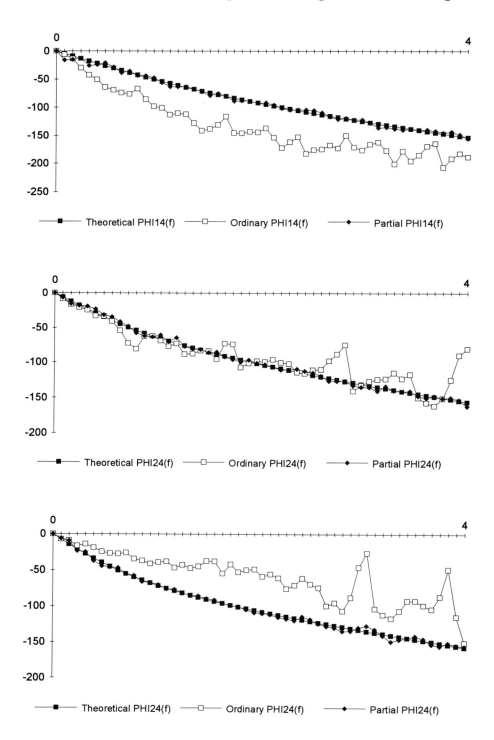

**Figure 7.11***b*  Transfer function phase estimates – correlated inputs

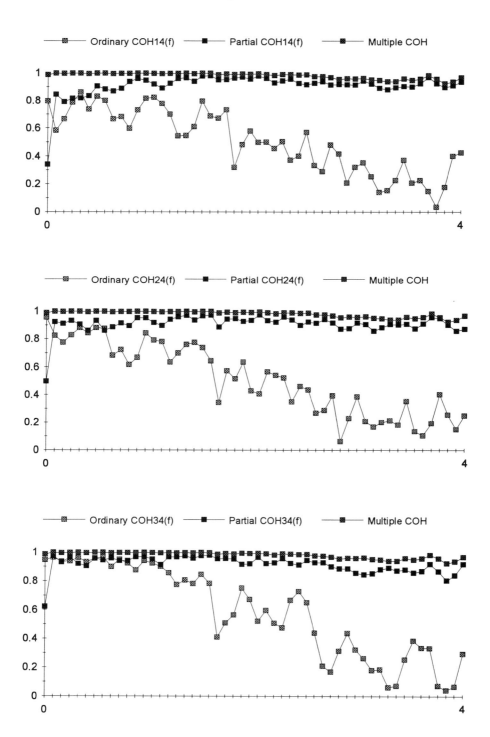

**Figure 7.11c** Coherence function estimates – correlated inputs

*7.5.2.3    Highly Correlated Inputs*

This is a *degenerate case* in which two or more of the inputs are *highly correlated* ($\gamma_{ij}^2(f) \approx 0.8$), as sometimes occurs in practice. Various scenarios can be simulated by setting the input-input gains $K_{12}$, $K_{13}$ and/or $K_{23}$ to higher values (say 10 or more). This produces degenerate partial transfer function estimates and/or spuriously high ordinary coherence estimates, and confirms the process model has redundant correlated input(s) that should be eliminated from the analysis. For example, if we set $K_{12}$ to a sufficiently high value such that there is a strong correlation (coherence) between inputs 1 and 2, we can simulate the effects of eliminating input 2 (say) in one of several ways:

1) *Zero Input Gain*: The gain for input 2 random series ($G_2$) is set to zero; that is, the random series is 'switched off'. In this case, however, there will still be 'leakage' from input 1 to the output if $K_{12} \neq 0$ and $K_{24} \neq 0$. Note that this 'leakage' effect yields a higher order transfer relationship between input 1 and the output, as shown by the large deviations in the transfer function modulus and, in particular, phase estimates.
2) *Zero Input Gain and $K_{24} = 0$*: Since $K_{12} \neq 0$, leakage may still occur between input 1 and input 3 via input 2 if $K_{23} \neq 0$. Note the changes in the input-output transfer function and coherence estimates.
3) *Zero Input Gain and $K_{12} = 0$*: This eliminates input 2 and reduces the order of the present MISO process model.

Ideally, however, this 'restructuring' of the MISO process model should be supported by *a priori* knowledge of the physical process wherever possible.

In the special case where *one* input has a dominant affect on the input-output transfer relationships such that the affects of the other inputs can be validly ignored, as assumed by equation (7.15), then the multivariate (MISO) process is reducible to a bivariate (SISO) process from an analysis viewpoint.

*7.5.2.4    Effects of Residual Output Noise*

The foregoing analysis results, Figures 7.10 and 7.11, were obtained with a small amount of *residual noise* added to the output ($G_4 = 1$, Table 7.1). However, as with bivariate processes, the magnitude of the residual noise can have a profound affect on the MISO transfer function and coherence function estimates, only more so. The reader may care to observe how these estimates change in the MISO simulation model as the gain of the output random series, $G_4$, is increased in magnitude.

## 7.6      MULTIVARIATE PARAMETRIC SPECTRAL ANALYSIS

The analysis of processes using multivariate parametric models have been extensively investigated in recent years in a wide range of disciplines, including geophysics (seismic data analysis), medicine (medical diagnostics), the process industries and the nuclear power industry (nuclear reactor surveillance and integrity monitoring). In the last case, for example, *extended* partial coherence ('EPCH') relationships based on autoregressive (AR) parametric models have been developed (e.g. Oguma 1982), and compared with those obtained using the non-parametric partial coherence ('PCH') results, equation (7.33), when applied to the multivariable noise analysis of nuclear power reactors. As expected from the discussion on closed loop processes given in Chapter 5, the EPCH method gives more reliable results than the PCH method when there are internal feedback loops present in the multivariate process, and both methods give equivalent results when there are no internal feedback loops.

A detailed account of multivariate parametric spectral methods is beyond the scope of this book. However, we note in passing that the benefits of AR methods for analysing multivariate closed loop processes may be coupled with the computational efficiency of the above recurrence algorithm, equation (7.42), by utilising the parametric power and cross-spectral methods discussed in Chapter 6, section 6.5, as shown in Figure 7.12. The *parametric* input-input and input-output power and cross-spectral density estimates $\hat{S}_{ij}(f)$ of the

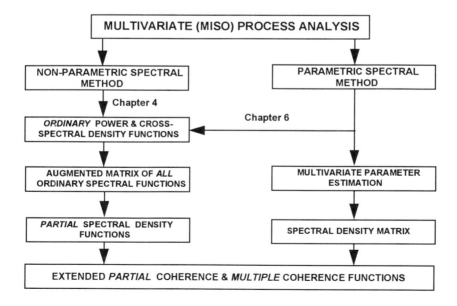

**Figure 7.12** Alternate multivariate parametric spectral methods

augmented spectral density matrix $\mathbf{S}_{NN}(f)$, equation (7.44), are computed using equations (6.77), (6.78) and (6.79). Once the elements of $\mathbf{S}_{NN}(f)$ are computed, the ensuing computational procedure is the same as to that discussed in section 7.4.

Multivariate parametric methods are a specialised field of study with numerous applications in a wide range of disciplines, and interested readers are referred to the literature in their field of study and the excellent text by Robinson (1983) for further details.

## 7.7     CONCLUDING REMARKS

Industrial processes are invariably multivariate in their dynamic behaviour, and, as we have seen in this Chapter, are difficult to analyse with any degree of certainty. As a general rule, the higher the order of the multivariate process, the more degrees of freedom, or 'flexibility', it can exhibit in its behaviour, and this, in turn, compounds the complexity of the analysis problem. This should be born in mind especially when attempting a 'black box' approach to the analysis problem; that is, where there may be very limited information about the process and one is forced to adopt a 'total ignorance' analysis approach in the initial experiment design stages. Having said this, however, if the underlying principles of this Chapter and the preceding Chapters are applied in conjunction with a comprehensive knowledge of the underlying processes involved, then consistent results will be achieved most of the time.

## REFERENCES

Bendat, J. S., 1976a, Solutions for the Multiple Input/Output Problem, *Journal of Sound and Vibration*, **44**, 311-325.

Bendat, J. S., 1976b, System Identification from Multiple Input/Output Data, *Journal of Sound and Vibration*, **49**, 293-308.

Dodds, C. J., & Robson, J. D., 1975, Partial Coherence in Multivariate Random Processes, *Journal of Sound and Vibration*, **42**, 243-249.

Faddeeva, V. N., 1959, *Computational Methods of Linear Algebra*, Dover Publications Inc., New York.

Granger, C. W. J., & Hatanaka, M., 1964, *Spectral Analysis of Economic Time Series*, Princeton University Press, New Jersey.

Jenkins, G. M., & Watts, D. G., 1968, *Spectral Analysis and its Applications*, Holden-Day, San Francisco.

Oguma, R., 1982, Extended Partial and Multiple Coherence Analyses and Their Application to Reactor Noise Investigation, *Journal of Nuclear Science and Technology*, **19**(7), 543-554.

Otnes, R. K., & Enochson, L. D., 1972, *Digital Time Series Analysis*, John Wiley and Sons Inc., New York.

Robinson, E. A., 1983, *Multichannel Time Series Analysis with Digital Computer Programs*, Goose Pond Press, Houston.

Romberg, T. M., 1978*a*, An Algorithm for the Multivariate Spectral Analysis of Linear Systems, *Journal of Sound and Vibration*, **59**(3), 395-404.

Romberg, T. M., 1978*b*, A Note on the Spectral Analysis of Linear Systems with Multiple Inputs and Outputs, *Journal of Sound and Vibration*, **60**(1), 149-150.

## TUTORIAL EXERCISES

1)  Prove that the matrix forms of $P_{kk:K}(f)$, $P_{kN:K}(f)$ and $P_{NN:K}(f)$, equations (7.24), (7.25) and (7.29), are given by:

$$P_{kk:K}(f) = S_{kk}(f) - \mathbf{S}_{k:K}^T(f)\mathbf{S}_{k:KK}^{-1}(f)\mathbf{S}_{k:K}(f)$$

$$P_{kN:K}(f) = S_{kN}(f) - \mathbf{S}_{k:K}^T(f)\mathbf{S}_{k:KK}^{-1}(f)\mathbf{S}_{N:K}(f)$$

$$P_{NN:K}(f) = S_{NN}(f) - \mathbf{S}_{N:K}^T(f)\mathbf{S}_{k:KK}^{-1}(f)\mathbf{S}_{N:K}(f)$$

2)  Derive the *theoretical* ordinary, partial and multiple coherence functions for a two-input single-output MISO process.

3)  Repeat Exercise 2, but for a three-input single-output MISO3 process.

4)  Write a computer program to compute the theoretical ordinary, partial and multiple coherence for Exercise 2 or 3, and by assuming appropriate values for the process parameters, compare your results with those computed with the MISO#SIM.EXE computer program:
    *a*)  independent inputs
    *b*)  correlated inputs
    *c*)  highly correlated inputs.
    What conclusions are you able to draw from your results?

5)  Repeat Exercise 4 but with *uncorrelated* white noise of variable amplitude added to the output. What conclusions are you able to draw from your results?

# APPENDIX A7

## MISO#SIM PROGRAM INSTALLATION AND OPERATION

MISO#SIM.EXE computes the partial spectral density functions, transfer functions and the ordinary, partial and multiple coherence functions for a three input single output MISO process. It has been designed to run under Windows 3.1 with video display settings of 640×480 pixels×256 colours, and also runs under Windows 95 with the same video display settings. Refer to the **readme.txt** document for further information.

## A7.1     INSTALLATION

1) Insert disc in the FDD and type **a:\setup filename** in the Windows 3.1 Program Manager File/RUN menu.
2) The **setup.bat** file copies *all* the files on the floppy disc to the hard drive directory **c:\filename**. (NOTE: The setup procedure will terminate if **filename** is not specified.)
3) The **setup.bat** file also decompresses the RUN library files **cable.dll** and **rlzrun20.rts** and instals them into a newly created Windows sub-directory **c:\windows\rlzrun20**.
4) For ease of operation, it is recommended that the four simulation programs in the **c:\filename** directory be installed as a program group icon with their separate program icons using the Windows Program Manager.

## A7.2     OPERATION

1) MISO#SIM.EXE can be operated in the normal manner using File Manager or as icons created with Program Manager.
2) Double click on the MISO#SIM.EXE file in File Manager or its icon in the Program Manager group window to **Run** the program. A "Title Form" is displayed and a **Run** option is added to the Menu Bar.
3) Program operation is straight forward and the data input options are discussed below. Choose the "Data File" **Default** button (for default data) then the **OK** button or **Enter** key to continue operation.
4) There is a waiting period as the program generates four (4) independent random series of data points (specified by NTotal), and a form with bar chart displays progress. When completed, a form with the first block of NData points of the input and output time series are displayed, then *tiled* forms with the input-output transfer functions and coherence functions are displayed.

5) **Ctrl+C** invokes a dialogue box which enables the user to terminate execution and return to the group window.

6) Use the **Run** menu options to execute a <u>N</u>ew Case or <u>E</u>xit from the program normally. Use the <u>F</u>ile Menu/<u>E</u>xit option to return to the group window.

7) The input-output theoretical, ordinary and partial transfer function moduli and phases, as well as the ordinary, partial and multiple coherence functions are printed to the Print Log for each case, and may be viewed using the <u>W</u>indow/Show <u>P</u>rint Log menu option. Use the <u>E</u>dit/<u>C</u>opy (Ctrl+Insert) menu option to copy the printed functions to Excel or other compatible spreadsheets for plotting (cf. Figures 7.10 and 7.11).

## A7.3    PROGRAM DATA INPUT

The default values of the program data input are displayed in the Data Input form as follows:

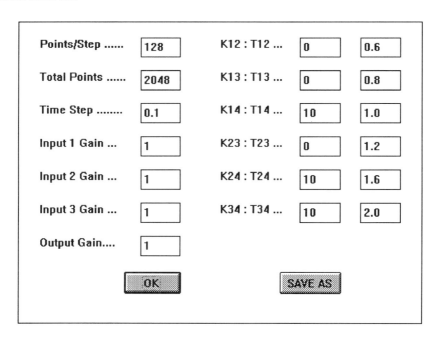

The program default data values are invoked by pressing the DEFAULT button in the Data File form. The data input values in the bordered text boxes may be altered for each new case (<u>R</u>un/<u>N</u>ew Case menu option) and saved as a **filename.dat** file by pressing the SAVE AS button and typing the **filename.dat** in the Save As form that is displayed. The saved file may be selected from the list given in the Data File form by pressing the OK button.

## A7.4     COMPUTATIONAL PROCEDURE

The program first generates the input and output random sequences of 2048 data values (NTotal), and then displays the process input and output time series for the first block of 128 data points (NBlock). (Note: for statistical reasons, the program checks the value of NTotal and sets NTotal = 4*NBlock if it less.) The theoretical and *ordinary* input-output transfer functions are then computed and displayed, and then the *ordinary* input-output coherence functions. The computations are repeated for each successive block of 128 data points, and the display forms are updated at the end of each computation step, which may be selected individually and minimized or maximized or, alternatively, viewed as a *cascaded* group (Window/Tile menu option). The MISO transfer function estimates are computed and displayed when the 2048 data values have been processed.

# 8

# PRACTICAL DATA ANALYSIS EXERCISES

## 8.1    INTRODUCTION AND LEARNING OBJECTIVES

In this Chapter, the reader is referred to the files labelled TEST#*n*.XLS ($n = 1,4$) on the companion disc included with this book. The disc data cover five different correlation and spectral analysis exercises, and are designed to give readers practical experience at computing some basic functions, including:

- the autocorrelation function $R_{xx}(\lambda)$, comparing the result with that of both a first order and a second order system whose responses are known exactly;
- the cross-correlation function $R_{xy}(\lambda)$ and its use to measure the average velocity of a flowing column of warm air;
- the power spectral density function $S_{XX}(\omega)$ and its use in detecting the onset of the anomalous operation of an industrial plant;
- comparison of the results with model solutions contained in the text.

The exercises are particularly helpful for implementation as part of a course on signal processing, or for use by practising professionals wishing to gain confidence in the use of the functions introduced in earlier Chapters. For those who do not wish to work through the exercises, fully worked solutions are given later in this Chapter.

Many of the exercises have been tested in engineering and information technology courses for undergraduates run by the authors over a number of years. Some of the exercises have been adapted from actual industry-based problems undertaken by the authors, and have been simplified to make laboratory-based exercises feasible within the time constraints that usually apply. Care has been taken, however, to ensure that the essential challenge of the problems remains unchanged. The reaction from the participants in our courses has been mostly positive, and they regard them as a very useful adjunct to the learning process by casting the theoretical applications in a practical setting.

## 8.2      ESTIMATION OF THE TIME CONSTANT OF A THERMOCOUPLE

In many practical situations thermocouples are used as the primary sensors in monitoring the condition of a major process variable in a plant. Often their use as diagnostic sensors of dynamic phenomena is an afterthought, long after the thermocouples have been installed, so a direct method of estimating the time constant of these sensors is not possible. It is necessary, therefore, to resort to using the random fluctuations in temperature inherent in the process as a calibration source and use the analysis of signals in an indirect way to estimate the time constant of the thermocouples. The method used is generic and can be applied to any first order system, as the following analysis demonstrates.

Consider the case of a thermocouple at an initial temperature $T_0$ suddenly immersed in a fluid of ambient temperature $T_m$. (This situation is relatively easy to approximate in the laboratory). Newton's law of heating and cooling states that the rate of change in the temperature is directly proportional to the temperature difference, and this may be written as:

$$\frac{dT}{dt} = k(T_\infty - T) \qquad\qquad \text{... (8.1)}$$

which is a first order differential equation solvable in many different ways. Note that the form of the solution of this simple equation is determined completely by one number $k$, the inverse of which is identified as the time constant $\tau$.

In general terms we may write the relationship between the input $x(t)$, to a first order system as characterised by a time constant $\tau$, to the output $y(t)$ by the following first order differential equation,

$$\tau \frac{dy(t)}{dt} + y(t) = x(t) \qquad\qquad \text{... (8.2)}$$

which, taking the Fourier transform of each side, yields

$$[j\omega\tau + 1]Y(\omega) = X(\omega) \qquad\qquad \text{... (8.3)}$$

Taking the complex conjugate of this relationship gives

$$[-j\omega\tau + 1]Y^*(\omega) = X^*(\omega). \qquad\qquad \text{... (8.4)}$$

Following the method outlined in Chapter 4, the relationship between the input and output power spectral densities is

$$S_{YY}(\omega) = \frac{S_{XX}(\omega)}{1+\omega^2\tau^2}. \qquad \qquad \text{... (8.5)}$$

Using the Wiener-Khinchin equations as defined in Chapter 2, the autocorrelation function can be estimated as

$$R_{YY}(\lambda) = \frac{1}{2\pi}\int_{-\infty}^{\infty}\frac{S_{XX}(\omega)e^{j\omega\lambda}d\omega}{[1+\omega^2\tau^2]}. \qquad \qquad \text{... (8.6)}$$

At this stage the assumption is made that the input temperature fluctuations have a white noise spectrum, which is a very common assumption made in many practical applications. It is especially necessary when, as in this case, only the output of the process is available for analysis. In practical terms, the assumption of a white noise input simply means that this applies over the frequency of interest.

The integration of the above requires the use of contour integration[1], recognising that the two poles of the integrand are given as

$$\omega = \pm\frac{j}{\tau} \qquad \qquad \text{... (8.7)}$$

and the result of the integration is given as

$$R_{YY}(\lambda) = \frac{1}{2\pi}2\pi j\sum\text{Residues in the upper half plane}, \qquad \text{... (8.8)}$$

which, on taking the input power spectral density as unity for convenience, gives

$$R_{YY}(\lambda) = \frac{1}{2\tau}e^{-\lambda/\tau}. \qquad \qquad \text{... (8.9)}$$

Thus, if the input power spectral density is white, a measurement of the autocorrelation function can be used to estimate the time constant of the system under test. This method of estimating the time constant of a thermocouple is used in many industrial situations where the thermocouple is already installed and therefore indirect methods must be employed. Despite its indirectness, the method is very easy to use, generally giving satisfactory results.

The reader may care to use the data stored as an MS Excel file Test#1.XLS on the disc to experiment with some data from a first order system using various sample lengths and averaging times in computing the autocorrelation function for the data given.

---

1.   See the note which summarises contour integration at the end of this Chapter.

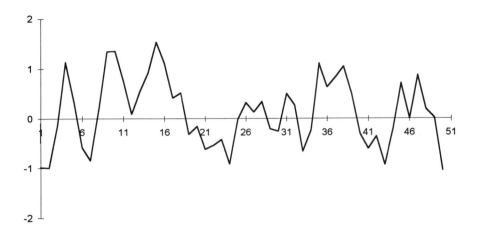

**Figure 8.1** Typical random fluctuations from a thermocouple (mean removed)

The data actually contain 5000 samples of the signal from a thermocouple sampled at equal intervals of time. A graph of the first 50 data points is shown in Figure 8.1. It is interesting to note that even with a relatively small sample of 2000, the exponential form of a first order system is clearly evidenced by the autocorrelation function shown in Figure 8.2. The reader may care to investigate the effect of longer sample lengths on the estimates of the time constant.

In Section 8.3, Chapter 8, the same methodology can be used to examine a second order system, again assuming a white noise input and that access is limited to the output.

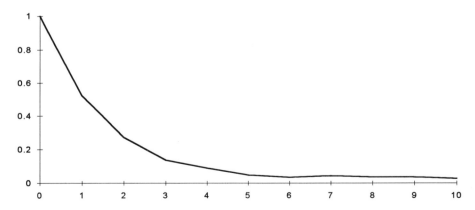

**Figure 8.2** Autocorrelation function of data in Figure 8.1 versus sample number averaged over 2000 samples

## 8.3 ANALYSIS OF A VIBRATING CANTILEVER

This exercise is derived from a project the authors completed a number of years ago on the assessment of the mechanical integrity of nuclear fuel elements operating in a water cooled reactor.

The fuel elements were suspended as vertically aligned cantilevers which were buffeted by the turbulent motion of the heavy water coolant. It was not possible to monitor the turbulence of the water and we could only access the top of each fuel element through direct mechanical connection via its sealing plug. Even with these severe constraints it was possible to make very detailed statements about the mechanical integrity of each fuel element.

It was necessary to assume that the input was the turbulence of the coolant water, which was assumed to have a white power spectral density, which forced the fuel elements into random vibration patterns. The vibration of each fuel element was detected by an accelerometer attached to the top of the sealing plug. All fuel elements were monitored in this way, producing a number of parallel outputs.

Although each fuel element was be monitored this way and, indeed, gross departure from 'normality' could be observed by visual examination of records of the random vibrations, subtle changes remained undetected by visual examination alone.

In order to establish a base for the early detection of small changes in the mechanical characteristics of any fuel element, it was necessary to adopt a standard against which departure from normality could be assessed. A theoretical standard was set by assuming that each fuel element behaved as a second order system excited by a white noise source. This assumption was made after observing the mechanical behaviour of other similar fuel elements under laboratory conditions where they could be excited by well known and carefully regulated forces.

The adoption of this theoretical model reduced the complexity of the analysis of the random signals to two parameters associated with the natural frequency of vibration and the damping coefficient of each cantilever respectively. Small variations in the internal mechanical structure of any fuel element were expected to be observable through either of these two parameters. The method consisted of estimating the autocorrelation function of the signal from each accelerometer, from which the natural frequency and damping ratio could be determined. This analysis method is precisely the same as the one on a first order system outlined in Section 8.2, but with slightly more complex algebra, using the following mathematical model.

The standard differential equation of a second order system in terms of the output of the system $y(t)$ (in this case the vibration level) and the input $x(t)$ (the turbulent excitation force provided by the coolant water) is

$$\frac{d^2 y(t)}{dt} + 2\xi\omega_n \frac{dy(t)}{dt} + \omega_n^2 y(t) = \omega_n^2 x(t). \qquad \dots (8.10)$$

Taking the Fourier transform of each side gives

$$((j\omega)^2 + 2\xi j\omega\omega_n + \omega_n^2).Y(\omega) = \omega_n^2 X(\omega) \qquad \dots (8.11)$$

and taking the conjugate of each side gives

$$((j\omega)^2 - 2\xi j\omega\omega_n + \omega_n^2).Y^*(\omega) = \omega_n^2 X^*(\omega). \qquad \dots (8.12)$$

Multiplying these last two equations gives

$$Y(\omega).Y^*(\omega) = \frac{\omega_n^4 X(\omega).X^*(\omega)}{(\omega^2 + 2\xi j\omega_n\omega - \omega_n^2)(\omega^2 - 2\xi j\omega_n\omega - \omega_n^2)} \qquad \dots (8.13)$$

and using the definition of power spectral density, results in

$$S_{yy}(\omega) = \frac{\omega_n^4 S_{xx}(\omega)}{(\omega^2 - \omega_n^2)^2 + 4\xi^2\omega_n^2\omega^2}. \qquad \dots (8.14)$$

Now, by using the Wiener-Khinchin equation we have

$$R_{yy}(\lambda) = \frac{1}{2\pi} \int_{-\infty}^{\infty} \frac{S_{xx}(\omega).\omega_n^4 e^{j\omega\lambda} d\omega}{(\omega^2 - \omega_n^2)^2 + 4\xi^2\omega_n^2\omega^2} \qquad \dots (8.15)$$

$$R_{yy}(\lambda) = \frac{1}{2\pi} \int_{-\infty}^{\infty} \frac{A.\omega_n^4 e^{j\omega\lambda} d\omega}{(\omega^2 + 2j\xi\omega_n\omega - \omega_n^2)(\omega^2 - 2j\xi\omega_n\omega - \omega_n^2)}. \qquad \dots (8.16)$$

Now the four poles are given by: $\pm\omega_n\sqrt{1-\xi^2} \pm j\xi\omega_n$ say $\pm\alpha \pm j\beta$, resulting in

$$R_{yy}(\lambda) = \frac{1}{2\pi} \int_{-\infty}^{\infty} \frac{A\omega_n^4 e^{j\omega\lambda} d\omega}{(\omega - [\alpha + j\beta])(\omega - [\alpha - j\beta])(\omega - [-\alpha + j\beta])(\omega - [-\alpha - j\beta])}$$

$$\dots (8.17)$$

The residue at

$$\omega = \alpha + j\beta \text{ is } \frac{A\omega_n^4 e^{j(\alpha+j\beta)\lambda}}{(2j\beta)(2\alpha)(2\alpha+2j\beta)}$$

and the residue at

$$\omega = -\alpha + j\beta \text{ is } \frac{A\omega_n^4 e^{j(-\alpha+j\beta)\lambda}}{(-2\alpha)(-2\alpha+2j\beta)(2j\beta)};$$

hence

$$R_{yy}(\lambda) = \frac{1}{2\pi} 2\pi j A\omega_n^4 \left[\frac{e^{j(\alpha+j\beta)\lambda}}{8\alpha\beta j(\alpha+j\beta)} + \frac{e^{j(-\alpha+j\beta)\lambda}}{8\alpha\beta j(\alpha-j\beta)}\right]$$

$$= \frac{A\omega_n^4 e^{-\beta\lambda}}{8\alpha\beta} \left[\frac{e^{j\alpha\lambda}}{\alpha+j\beta} + \frac{e^{-j\alpha\lambda}}{\alpha-j\beta}\right] \qquad \text{... (8.18)}$$

Substituting for $e^{j\alpha\lambda} = \cos(\alpha\lambda) + j\sin(\alpha\lambda)$ and $e^{-j\alpha\lambda} = \cos(\alpha\lambda) - j\sin(\alpha\lambda)$, (8.18) then becomes:

$$R_{yy}(\lambda) = \frac{A\omega_n^4 e^{-\beta\lambda}}{8\alpha\beta} \left[\frac{2\alpha\cos(\alpha\lambda) + 2\beta\sin(\alpha\lambda)}{\alpha^2 + \beta^2}\right]$$

$$= \frac{A\omega_n^4 e^{-\beta\lambda}}{8\alpha\beta\sqrt{\alpha^2 + \beta^2}} \cos(\alpha\lambda + \phi)$$

where $\phi = \tan^{-1}\dfrac{\alpha}{\beta}$. This equation can be written in damped harmonic form as:

$$R_{yy}(\lambda) = \frac{A\omega_n e^{-\omega_n\xi\lambda}}{\xi\sqrt{1-\xi^2}} \cos(\omega_n\sqrt{1-\xi^2}\lambda + \phi). \qquad \text{... (8.19)}$$

where $\alpha = \omega_n\sqrt{1-\xi^2}$ and $\beta = \omega_n\xi$.

Note that the autocorrelation function is an exponentially damped cosine wave and that the natural frequency $\omega_n$ and the damping ratio $\xi$ can be estimated by conventional methods from a plot of this function. [2]

The laboratory exercise, derived from the above experience, utilises a

---

2. It is worth noting that as the damping ratio tends to zero, the value of the autocorrelation function tends to infinity. In practice an extrapolation of a plot of the inverse of the value of the autocorrelation function at zero lag can indicate the conditions under which dynamic instability may occur.

vertical wooden cantilever excited into random vibration by the impact of a turbulent jet of air. The vibration is measured by an accelerometer clamped to the fixed end of the beam.

In the courses which we taught, students were required to derive the mathematical model outlined above to help them understand the use of Wiener-Khinchin equations and sharpen their skills in contour integration. In addition, the laboratory exercise required the students to consider the choice of sample rate and sample length necessary to produce reliable estimates of the autocorrelation function. Students never failed to be impressed by the fact that they could extract useful information by the application of correlation analysis to random data that appeared to contain nothing of any utility when examined visually. They always found that the assumption that the system was of second order was somewhat imprecise but, even so, they appreciated the value of the method as a potential industrial diagnostic tool.

Data taken from the laboratory exercise are provided on the disc as Test#2.XLS and typical results that may be used for comparison purposes are given in Figures 8.3 and 8.4.

It should be emphasised that the assumption that a cantilever can be analysed as a second order system is just that, *an assumption*, albeit an intuitively plausible one. The value of this approach is that it reduces the complexity of the information in the random signal shown in Figure 8.3 to a situation where the cantilever's condition is summarised by two numbers only. In practice these numbers would be logged, and action initiated if significant movements in these parameters occurred.

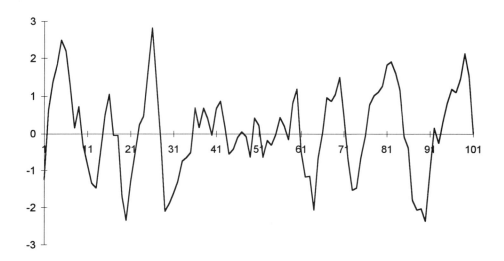

**Figure 8.3** Typical sample from a randomly vibrating cantilever

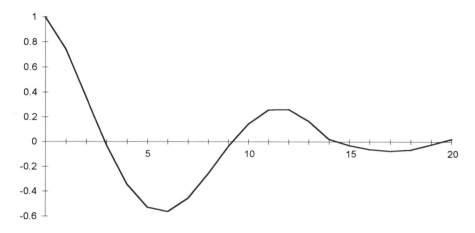

**Figure 8.4** Autocorrelation function of data from a randomly vibrating cantilever shown over 20 data samples

## 8.4    MEASUREMENT OF AVERAGE VELOCITY AND MIXING LENGTH IN TURBULENT FLOW

In industry it is often necessary to measure the average velocity of a moving, continuous solid, liquid or gas. In many applications involving a gas or a liquid, well tried methods such as turbine flow meters, orifice plates or rotameters are used where the flowing substance presents little if any difficulties with these methods. On the other hand, in some situations the environment is such as to preclude these techniques and other methods must be found. One such method is the use of cross-correlation analysis between the output of two robust sensors separated a known distance along the direction of the flow, the sensors being responsive to some physical parameter which is transported along with the flow. Typical examples would be temperature or conductivity within the fluid which cause fluctuations in the output from each sensor as the fluid passes by them. This situation is shown diagrammatically in Figure 8.5.

It is usual to discard the mean value of the signals from each of the sensors and develop a suitable signal processing strategy around the analysis of the fluctuations $x(t)$ and $y(t)$ by assuming that they are related. It is recognised that in an ideal situation the signal from the downstream sensor is simply a delayed version of that from the upstream sensor. In practice, turbulence or the injection of other noise sources render this assumption invalid, and a more realistic model is to use the following expression:

$$y(t) = \xi x(t - \tau) + n(t).$$

where $\xi \le 1$ and $\tau$ is the transport delay.

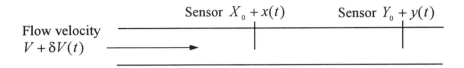

**Figure 8.5** Illustrating the concept of flow measurement by cross-correlation.

The addition of the noise source $n(t)$ is an attempt to account for the presence of uncorrelated noise sources. With this model it is easy to show that the cross-correlation function between the signals from the two sensors reaches its maximum when the computational delay equals the transport delay.

When introducing students to the practical application of cross-correlation analysis to measure transport delays we have found it beneficial to start with simple, well defined simulations on a computer. This has the real advantage that transport delay and signal-to-noise ratio can be set in precise terms. It does not, as we shall see later, include some of the complexities inherent in real life.

A diagrammatical representation of a simulation is shown in Figure 8.6. The random input signal and the uncorrelated noise source are easily generated by random number generators found in most spreadsheet programs. The output signal is the sum of the delayed signal and the noise source. This output signal is correlated with the input signal at various signal-to-noise ratios to recover the actual value of the delay. In this case the delay is set at ten data samples.

Typical results of this exercise are shown in Figures 8.7a to d for a range of signal-to-noise ratios covering the range from 0 to −20 dB.

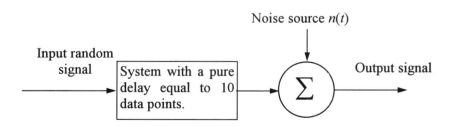

**Figure 8.6** Simulation used to introduce time delay estimation by cross-correlation analysis

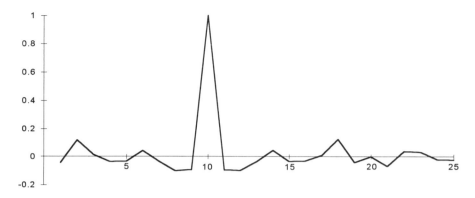

**Figure 8.7***a* Cross-correlation function versus sample number (no added noise)

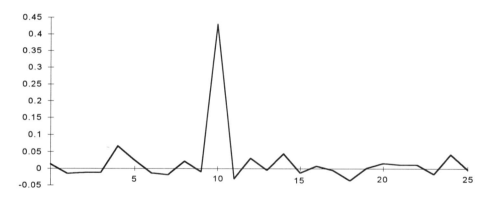

**Figure 8.7***b* Cross-correlation function versus sample number for −6dB S/N ratio

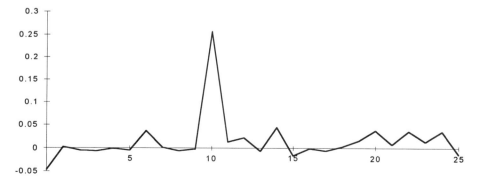

**Figure 8.7***c* Cross-correlation function versus sample number for −12dB S/N ratio

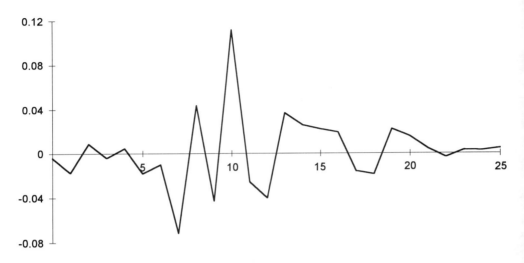

**Figure 8.**7*d* Cross-correlation function versus sample number for –20dB S/N ratio

Note that the correlation function decreases from a value of 0.7 when the signal-to-noise ratio is 0, to around 0.08 at –20 dB signal-to-noise ratio. It is only possible to be confident when using the latter figure because it is expected that there will be a transport delay of precisely ten data points. This situation would occur in practice if a number of sensors spaced at known locations along the flow were used to estimate the correlation coefficient between pairs of sensors with ever-increasing distances between them. This aspect is covered in the following exercise.

One laboratory-based exercise which challenges (and some times frustrates) students is one in which hot air from a domestic hair drier is ducted down a perspex tube along which a series of thermocouples are installed. The thermocouples, mounted normal to the flow, detect the mean temperature of the hot air and the small random fluctuations superimposed upon it. The object is to use the thermocouples in pairs with ever-increasing separations in an attempt to measure the transport delay and the steady decline in the correlation coefficient as the thermocouples are spaced further apart, the latter measurement having relevance in commenting on the efficiency of the mixing taking place in the fluid. This 'mixing length' has application in heat transfer studies.

The data for this exercise is in Test#3.XLS on the disc, and this records signals from a pair of thermocouples spaced 25 mm apart. The thermocouples were deliberately unshielded in order to make the exercise include some interference from the mains frequency. In an industrial situation, of course, every attempt would be made to minimise such mains interference. Typical results are recorded in the Figures 8.8, 8.9 and 8.10.

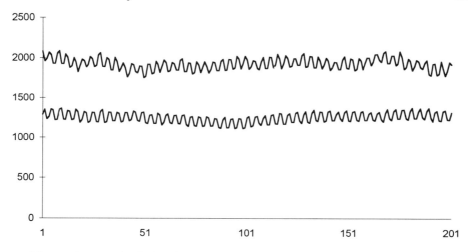

**Figure 8.8** Typical signals from two thermocouples spaced 25 mm apart

The signals from the thermocouples shown in Figure 8.8 are clearly contaminated by a sinusoidal disturbance observable on both channels and attributable to mains interference. It is relatively easy to filter out this disturbance by using a low pass-filter and at the same time remove the mean value from each signal to produce the result shown in Figure 8.9.

It now remains to estimate the transport delay by cross-correlating the signals as shown in Figure 8.9 to produce the result shown in Figure 8.10. The upper curve is obtained by delaying the up-steam sensor with respect to the down-stream one. The lower curve is with the down-stream sensor delayed. It is good practice to compute correlograms with positive and negative lag to check that the sensors are correctly identified.

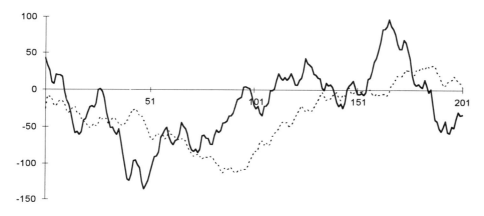

**Figure 8.9** Signals from the thermocouples, filtered and with means removed

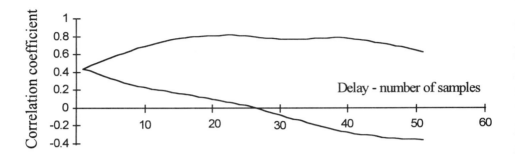

**Figure 8.10**  Typical cross-correlogram between the two thermocouple signals showing positive lag (upper) and negative lag (lower)

Note from Figure 8.10 that the transport delay is equivalent to the time passage of 20 data points and thus can be estimated if the sample rate is known, which is usually the case in industrial applications.

## 8.5      LOCATION OF A NOISE SOURCE IN A DISPERSIVE MEDIUM

In this section the slightly more difficult problem of the cross-correlation of signals propagated in a dispersive media is examined. Such a situation would arise, for example, in the transmission of elastic waves through any solid medium. A specific example would be the detection of a leak in a solid pipe by the correlation of acoustic signals which travel along the pipe, the acoustic signals resulting from the noise produced by the escape of the fluid from the pipe into the surrounding environment.

Although the signals from any leaking pipe are usually very easy to detect in practice, using accelerometers clamped to the pipe, for example, considerable care must be exercised in the use of cross-correlation calculations. In this case, as the medium is dispersive, the induced shift in the frequency spectrum results in the straightforward calculation of the cross correlation coefficient producing disappointing results.

The data shown in Figure 8.11 is from a simulation where two sensors were spaced a distance apart, which was equivalent to 20 data points of delay. It is just possible to estimate this by visual inspection, although not with a great deal of confidence. These data are provided in disc number one as file Test#4.XLS.

Given the clean-looking signals portrayed in Figure 8.11, it is natural to try to estimate the time delay between them by direct application of cross-correlation analysis. Remember, though, that in this case the medium through which the signals are propagated is dispersive. As a result, the frequency spectrum shifts as the energy moves along the transmission path with a very noticeable

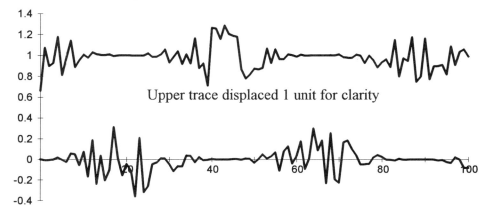

**Figure 8.11** 100 data points of the raw signals from two sensors spaced the equivalent of 20 data points apart

deleterious effect on the estimate of the cross-correlation coefficient, as demonstrated in the Figure 8.12.

Note the general low correlation coefficient which has a maximum around 0.2. Note particularly that the indication of the delay between the two signals is incorrect.

The signal processing strategy needed in this case is one which detects the envelope of the raw signal. The impulsive bursts of acoustic energy from the leak form the envelope which modulates the underlying high frequency acoustic noise. Other measures such as the mean-square or the absolute value are reasonable approximations for use of envelope detection.

The result of using the absolute value in a cross-correlation analysis is shown in Figure 8.13. Note that the greatly improved correlation coefficient is about 0.7 and that the delay is indicated at the correct number of sample points.

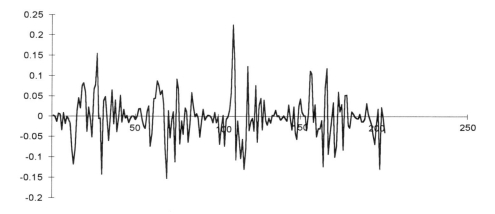

**Figure 8.12** Cross-correlation coefficient for the data shown in Figure 8.11

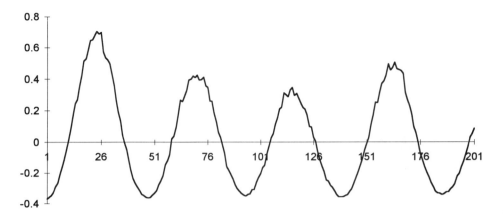

**Figure 8.13** Cross-correlation coefficient of the absolute values of the raw data
shown in Figure 8.11 over a delay of 200 samples

## 8.6     DETECTION OF A SODIUM-TO-WATER LEAK IN A HEAT EXCHANGER

This is the most challenging exercise presented in this Chapter. It is based on
the need to detect an incipient sodium-to-water leak in the heat exchanger of a
liquid metal cooled fast reactor. If the leak can be detected at an early stage,
then remedial action can be taken to limit its potential mechanical damage to
the plant.

The data presented for this exercise are a sample taken from recent research
in the area, and they contain signals of the background noise from the prototype
fast reactor at Dounreay in Scotland mixed with actual sodium-to-water
injection tests results from the USSR. The researchers were presented with a
number of test situations with a variety of signal-to-noise ratios with a signal
bandwidth of 40 kHz and a sample rate of 160 kHz. Their aim was to detect the
start of the sodium-to-water injection and its duration.

In designing an assignment for undergraduate students, a representative
sample of data was selected from that provided to all researchers participating
in the project. These data are provided on computer disc two as a file of
600,000 short integers. The injection of the sodium-to-water appears after an
unknown number of samples from the start of the file and continues for an
unknown duration. The exercise is to detect the onset and duration of the
sodium-to-water reaction by using an appropriate signal processing strategy.

Most students completed this exercise reasonably well. One student
produced excellent results by painstaking attention to detail, and he located the

existence of a very narrowband process[3] using a band-pass filter only 30 Hz wide. A typical result from his assignment is shown in Figure 8.12.

Some guidelines on one method that could be used for this exercise are outlined below. This method is based on the use of very narrowband filtering in the spectral domain, and relies on knowing the likely spectral composition associated with a leak.

### Choice of sample length

One way to establish when the sample length is sufficient to ensure the process can be considered to be statistically stationary is to compute some statistical measure over various sample lengths and note when that estimate remains reasonably steady. In this case mean value remained constant after about 4000 samples. It is useful to use the mean in this part of the exercise, as it is usual to reduce the signal to one of zero mean value for subsequent analysis.

### Feature extraction and establishment of a standard reference

As indicated in Section 8.3, the early detection of the onset of an anomaly is a common feature of industrial diagnostics, and this is highlighted in this exercise. The choice of an appropriate feature on which to base the decision rules is the key to successful implementation of a diagnostic scheme. Wherever possible it is advantageous to use any *a priori* knowledge that may be available. In this case it is well known that a sodium-to-water reaction will, among other things, produce many bubbles of hydrogen and oxygen  Any gas bubble in a liquid will act as a very efficient source of acoustic noise of almost pure tone, determined entirely by the radius of the bubble. Although it is not possible to predict the precise size of bubbles produced in a sodium-to-water reaction, laboratory work indicates that they will generally be of a size that produces acoustic signals around 0.5 to 2 kHz. Some form of narrowband pass filter would therefore seem to be appropriate.

As it is known that the sodium-to-water injection does not take place until part way through the process, the value of any extracted feature at the start of the record can be used as the base line standard against which any departures can be assessed. The use of an averaging time in the region of 4000 samples of data means that the output of any feature extraction strategy can be updated every 4000 data samples, or at about every 0.66% of the total length of the available data.

---

3.    Although fault location by use of narrow band detection is valid when the signal to be detected is well known, there are considerable reservations with this approach. Some alternative generic methods of incipient fault location are discussed in Case Study C.

**Checking the hypothesis of a narrowband process**

In order to allow for the fact that the acoustic noise caused by the sodium-to-water injection is likely to be spasmodic, it is necessary to compute the power spectral density over a relatively large number of samples to average out the signal over a number of bursts of acoustic energy. Gerard Carter found that calculating the average of ten spectral estimates, each using 8192 samples, at the start of the record and again at the end of the record clearly indicated the existence of a very narrowband process at the end of the file. This narrowband signal, centred on $1113^4$ Hz, was not evident at the start of the file. The output of a band pass filter with cut-off frequencies $30^5$ Hz either side of the frequency detected above was thus chosen as the feature to be monitored throughout the files.

**Design of the bandpass filter**

To complete this section the basic steps are outlined for designing a suitable Finite Impulse Response Filter (FIR) to perform the task in hand. The ideal response of the FIR filter is shown in Figure 8.14 as a rectangular-shaped filter which can be modelled by an infinite Fourier series. In practice, however, the Fourier series is truncated after a finite number of terms usually denoted by $M$, with a narrowband filter requiring a large value of $M$. The Fourier series for $H(e^{j\omega_n})$ is given as

$$H(e^{j\omega}) = \sum_{m=-\infty}^{\infty} h_m e^{+j\omega m}$$

$$h_m = \frac{1}{2\pi} \int H(e^{j\omega}) e^{-j\omega m} ;$$

and, therefore, the required Fourier series coefficients are

$$h_m = \frac{1}{70}\text{sinc}\left(\frac{\pi m}{70}\right) - \frac{1}{74}\text{sinc}\left(\frac{\pi m}{74}\right).$$

The required filter expressed in the $z$ domain together with its 'window' is

$$H(z) = z^{-M} \sum_{m=-M}^{M} h_m w_m z^{-m} .$$

---

4.    This value was established by a student as part of a set assignment set using the data of Test#4 for a BEng project. His result is shown in Figure 8.12.
5.    This choice results in the upper and lower cut-off frequencies being $\pi/70$ and $\pi/74$, respectively, with the sample rate of 160 kHz being scaled as $2\pi$.

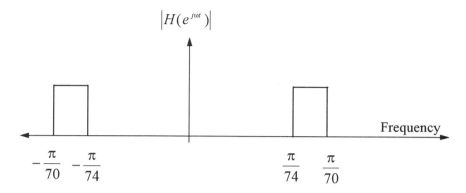

**Figure 8.14** Illustration of the choice of bandpass filter

The window coefficients $w_m$ may be any of the ones in common usage. A typical result for a Blackman filter with $M = 1000$ is shown in Figure 8.15.

A window is necessary to reduce the ripples in the amplitude response of the filter due to using only a finite value for the parameter $M$. In this case the window coefficients are given as:

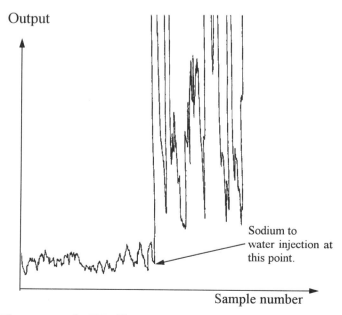

**Figure 8.15** The output of a FIR filter using a Blackman window as a function of sample number. (Note incipience of the sodium-to-water leak.)

$$w_m = 0.42 + 0.5\cos\left(\frac{\pi m}{M}\right) + 0.08\cos\left(\frac{2\pi m}{M}\right) \text{ for } m \le M$$

$w_m = 0$ otherwise.

## NOTE ON CONTOUR INTEGRATION

(For use with reference to Sections 8.2 and 8.3, Chapter 8)

A suitable starting point is to note that the integral around any closed contour for an analytical function (i.e. a function that is differentiable everywhere within the contour) is always zero. If, for example, a function $\Phi(z)$ contains one simple pole at $z = a$ then we may write the function in terms of an analytic function thus:

$$\Phi(z) = \frac{F(z)}{[z-a]}$$

The integration of the above function from $-\infty \to +\infty$ is usually done by choosing a suitable contour as shown in Figure 8.16. This choice ensures that the function is analytic within it, resulting in:

$$\oint \Phi(z)dz = 0.$$

The contour consists of six paths with two of them cancelling each other. In engineering problems it is usual to deal with functions that are zero at infinity, and hence the contour integral reduces to the following:

$$\oint \Phi(z)dz = \int_{-\infty}^{+\infty} \Phi(z)dz + \int_{P_3} \Phi(z)dz = 0.$$

To evaluate the integral around $P_3$ we write $[z-a] = \varepsilon e^{j\theta}$ which gives

$$\int_{P_3} \frac{F(z)dz}{[z-a]} = \int_0^{2\pi} \frac{F(a + \varepsilon e^{j\theta})j\varepsilon e^{j\theta}d\theta}{\varepsilon e^{j\theta}}.$$

It can be seen that the exponential terms cancel, resulting in the following expression as the radius of the circle shrinks to zero.

$$\int_{-\infty}^{\infty} \frac{F(z)dz}{[z-a]} = 2\pi j F(a)$$

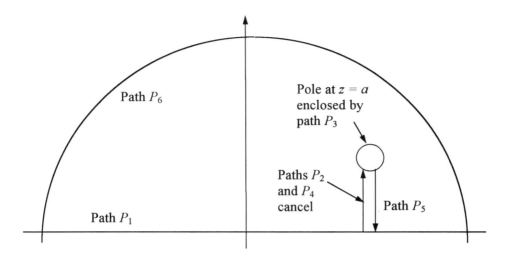

Pole at $z = a$
enclosed by
path $P_3$

Path $P_6$

Paths $P_2$
and $P_4$
cancel

Path $P_5$

Path $P_1$

**Figure 8.16** Illustrating the contour used for integration

Similarly :
$$\int_{-\infty}^{\infty} \frac{F(z)dz}{[(z-a)(z-b)]} = 2\pi j \left\{ \frac{F(a)}{(a-b)} + \frac{F(b)}{(b-a)} \right\}$$

The right hand side is written as $2\pi j \sum [\text{residues in the upper half plane}]$.

# PART TWO

# INDUSTRIAL
# CASE
# STUDIES

# Case Study A

## DETECTION OF A SODIUM-TO-WATER LEAK IN THE HEAT EXCHANGER OF A LIQUID METAL FAST BREEDER REACTOR

This case study is an edited version of a paper by S. E. F. Rofe and T. J. Ledwidge presented to a research coordination meeting of the International Atomic Energy Agency held in Kalpakkam, India, 1-3 November 1994

### A.1 CONTEXT

The genesis of all nuclear power programmes operating by atomic fission relies on the existence of a naturally occurring fissile isotope $U^{235}$ which comprises 0.7% of natural uranium. Natural uranium is used directly in a number of different nuclear reactors where the neutrons produced by the fission process are moderated by substances such as graphite or heavy water in order to slow them down to velocities at which capture by a $U^{235}$ atom is probable, resulting in a further fissionable event. These reactors are called *thermal reactors,* or *thermal converters* if their role is to convert natural uranium to fissionable plutonium.

In a *fast reactor* the amount of $U^{235}$ is increased to levels considerably above 0.7%, resulting in a high probability of fission, with un-moderated neutrons travelling at very high velocities. Alternatively, fast reactors may use plutonium produced in the thermal converter programme. In a fast reactor the use of neutrons is very economical because there is none of the energy loss which accompanies the moderation process resulting in an intense power density when compared to that of a thermal reactor.

The high power density in fast reactors requires that the heat removal process needs a heat transfer medium which is capable of dealing with both elevated temperatures and a high heat flux, so liquid sodium or a sodium-potassium alloy are generally used to convect the heat from the reactor core.

The heat from the liquid metal is utilised by passing it through a heat exchanger where it converts water to steam. The steam generated exits from the heat exchanger and drives a steam turbine to generate electricity in the conventional way.

The demands on the mechanical integrity of the heat exchanger are severe, because not only does it have to deal with the corrosive nature of the liquid metal, but it must also handle the possibility of a sodium-to-water reaction in the event that a leak occurs in any of the heat exchanger's tubes through which the water is pumped. Even a small leak in one tube may grow in size and the steam escaping at high velocity from the ruptured tube may cause further damage to tubes in the neighbourhood.

The need for an industrial diagnostic system to detect the occurrence of a small leak before it escalates to cause further damage is thus a high priority in all fast reactor heat exchanger installations, and is currently the subject of substantial research around the world.

Significant acoustic noise is generated in a heat exchanger when one of its steam tubes develops a leak. There are three major sources of acoustic noise of varying spectral composition.

1. The noise due to an expanding jet of steam escaping from the site of a leak is generally regarded as a broad band acoustic signal that will cause parts of the structure of the heat exchanger to vibrate in sympathy.
2. The noise due to droplets of liquid metal convected against a neighbouring tube by the high velocity steam jet is believed to be a narrowband process centred around 0.5-1 MHz.
3. The oscillation of the bubbles of hydrogen or oxygen produced by the sodium-to-water reaction produce spectra in the low kHz region.

In principle it is possible to detect all these acoustic signals through mechanical waveguides attached to the outside of the heat exchanger vessel. Waveguides are terminated in conventional accelerometers or other piezo-electric devices which convert the mechanical energy into electrical signals for further analysis.

## A.2     SOURCE OF THE TEST DATA

This case study is adapted from part of a coordinated research programme sponsored by the International Atomic Energy Agency in Vienna. It is a benchmark test using data on the background noise from the Prototype Fast Reactor at Dounreay in the north of Scotland, mixed with signals from an experimental 3.8 gms/s water injection into sodium carried out in an acoustic rig in the USSR. The composite data were mixed together and provided in digital format by the CEA, Cadarache, France.

The data consisted of seven sets of four files recording the outputs of four acoustic sensors distributed around the experimental rig. One set contained background noise only, one set of leak noise only, and the five remaining sets contained mixed test data at five different signal-to-noise ratios varying from −6dB to −24dB. Each of the test files was 7.8 seconds long and contained signals from a leak lasting in excess of 3.8 seconds long. The analogue to digital conversion was at 131,072 Hz with a useful frequency bandwidth of 51.2 kHz.

The major task was to detect the onset and duration of the leak from each of the five signal-to-noise ratios used in the benchmark test. A secondary, but no less important, task was to attempt to locate the spatial coordinates of the leak and determine if the use of multi-track data provided any significant advantage over the use of single track data used in previous bench mark tests.

Although access to data from background and leak signals determined separately provided an opportunity to design a fault detection strategy around these known characteristics, the authors used a generic approach that did not rely on this specific *a priori* information.

This philosophical stance recognised that small variations from experimental conditions generated in a test rig, compared to those that apply in a real operating plant may significantly change the signal characteristics. Here the reader should compare the alternative strategy used in the practical data exercise given in section 8.6 of Chapter 8, where the emphasis was on the use of spectral filtering techniques to extract the onset of an anomalous condition. In this exercise the detection of an anomaly was accomplished under very adverse signal-to-noise ratios, but only by exploiting the spectral characteristics of the anomalous event determined from test data supplied to the experimentalist. In many practical situations it is not possible to generate test data from a real or simulated anomalous event and thus the use of generic detection strategies becomes the only possible method.

## A.3     SIGNAL MODEL

An appropriate signal model is shown diagrammatically in Figure A..1. Here some   assumptions are necessary. Firstly, the background noise is recorded equally by all four sensors, while the signals from the leak are taken to be dependent on the position of the leak. It should be noted that the gas and liquid mixture surrounding the site of the leak forms a **highly dispersive** medium through which part of the acoustic energy propagates. Other acoustic energy travels through the mechanical structure of the test rig which is a **less dispersive** medium.

These effects are represented in Figure A1 by the convolution of the signal from the leak with four different linear filters $\{h_i(t): i = 0 \rightarrow 3\}$. The different transport delays between the signal source $x(t)$ and each of the sensors is accounted for by the introduction of four values of the lag term $\{\lambda_i: i = 0 \rightarrow 3\}$. The four outputs from the sensors are thus the additions of the outputs from the linear filters to the uncorrelated noise sources.

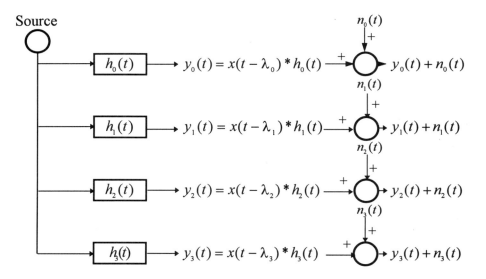

**Figure A.1**  Assumed signal flow model

### A.3.1    Power Spectral Density of the Background Noise.

Spectral estimates of the background noise were determined using a periodogram approach. These estimates averaged 4096 point FFTs (overlapping by 2048 samples) over the entire files for each of the four sensors. Typical results for sensor one and sensor four, shown in Figures A.2 and A.3, indicate that the background spectra are very similar; all four sensors show the same results up to 16 kHz with minor variations of the peaks of the harmonics above that frequency. These findings lend support to the assumption that the background noise is detected equally by all four sensors.

### A.3.2    Power Spectral Density of the Signal from the Leak

Spectral estimates of the signals from the leak were determined in precisely the same way as those for the background noise. Typical results for two of the four sensors are shown in Figures A.4 and A.5.

Note that these spectra are very different from those of the background noise and also very different from each other. As the sensors and their associated communication channels used were identical, this observation suggests that the path from the leak site to the sensors modifies the characteristics of the signals significantly.

Given the spectra of the background noise and the signals from the leak, it is possible to design a feature extraction strategy to detect the onset and duration of the leak by using appropriate narrowband filters on each channel. However, this approach was rejected in favour of developing a generic approach.

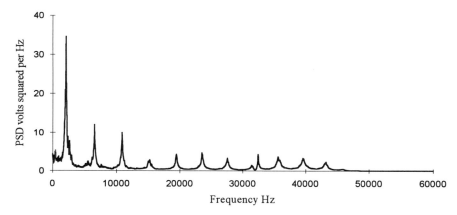

**Figure A.2** Power spectral density of the background noise as detected by sensor 1

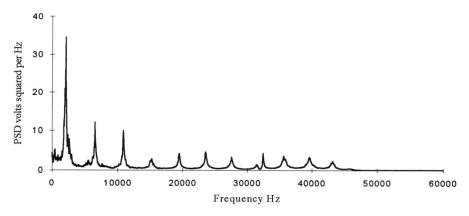

**Figure A.3** Power spectrum density of the background noise as detected by sensor 4

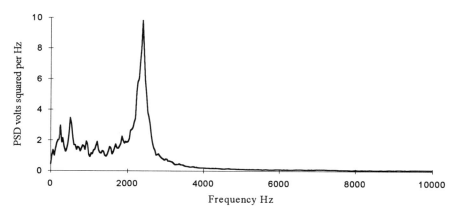

**Figure A.4** Power spectral density of the signal from the leak, sensor 1

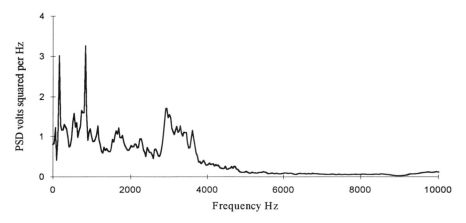

**Figure A.5**  Power spectral density of the signal from the leak, sensor 4

## A.4       AR MODELLING

Autoregressive (AR) modelling has been developed extensively over a number
of years for modelling linear stochastic processes. The AR model consists of a
set of parameters which can be used for either prediction or spectral estimation.
The $p$th order AR linear predictor is given as

$$x[j] = -\sum_{k=1}^{p} a[k]x[j-k] + u[j],$$

where $a[k]$ are the AR parameters and $u[n]$ is a zero mean white Gaussian
distributed process with variance $\sigma^2$. The spectral estimate is given as

$$P_{AR}(f) = \frac{\sigma^2}{1 + \sum_{k=1}^{p} a[k]e^{-j2\pi kf}}.$$

There are several methods which may be used to determine the parameters of
the AR model for example, Kay and Marple (1981), Marple (1987) and Kay
(1988). Of these methods, the modified covariance method minimises the
average of both the forward and backward prediction errors, as opposed to the
autocorrelation or covariance method which minimises the forward prediction
error only.

The two variables which must be set in determining the AR model are the
order $p$ and the number of samples used in the estimate.

## A.4.1    Model Order Selection

The problem of model order selection has been dealt with in considerable detail by Akaike (1970), Rissanen (1983), Kay and Marple (1981), Marple (1987), and Kay (1988). In general, if the model order is too low, the result will be a high prediction error and a smoothed spectrum. Conversely, if the model order is too high, spectral peaks will result in an attempt to model the noise process (Kay 1988). Optimal criteria have been developed by Akaike and Rissanen independently. Akaike developed the final prediction error (FPE) and the Akaike information criterion (AIC), whereas Rissanen developed the minimum description length (MDL) criterion. These are defined as follows:

$$FPE(p) = \frac{N+p}{N-p} \tilde{\rho}_p$$

$$AIC(p) = \log \tilde{\rho}_p + \frac{2p}{N}$$

$$MDL(p) = \log \tilde{\rho}_p + \frac{2p}{N} \log N$$

where $\tilde{\rho}_p$ is the error variance estimate for model order $p$, and $N$ is the number of samples used for the estimate. Sometimes a minimum can be found when applying these criteria, but often the criteria are simply monotonically decreasing. In that case the choice of the model order is a question of judgement; a value of $p$ being chosen such that the AIC shows insignificant decreases above that value. In this case study, all three criteria gave very similar results, that for the AIC criteria being shown in Figure A.6. As Figure A.6 showed that the AIC decreased rapidly initially and levels off around a model order of 32, this model order was used for all subsequent analysis.

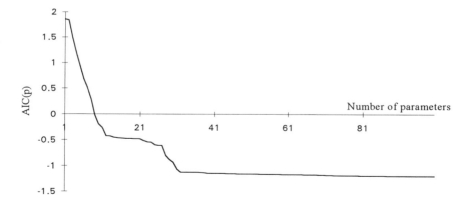

**Figure A.6** AIC criterion for the background noise

### A.4.2    Data Length

As indicated in Chapter 1, a key parameter in the decision making process is the time taken to recognise a fault condition. This decision time is directly related to the number of samples used in any estimation scheme. Examination of the variation of each of the 32 AR parameters showed that, for the autocorrelation method, as the sample size was increased, no convergence occurred over the maximum length of sample used, indicating that this method was not suitable for the context. Conversely, the modified covariance method resulted in only gradual change to the parameters after some 20,000 samples This length of sample implies an effective averaging time of 153 ms with a sampling frequency of 131,072 Hz. Subsequently, all analysis was computed with a 32nd order model using 20,000 samples.

## A.5    FAULT DETECTION STRATEGIES

The underlying assumption with all fault detection strategies that employ any form of modelling is that the occurrence of a fault changes the model structure of the received signal. A direct comparison between parameter estimates obtained under 'normal' operating conditions and those obtained during any further operational conditions may indicate the onset of an anomaly. Two of the different detection strategies used in this case study are explained in the following sections. A less effective method of likelihood ratio tests was reported in our 1994 paper but is omitted here for the sake of brevity.

### A.5.1    Mean square of prediction error

The AR model of order $p$ gives the parameters of the $p$th order linear predictor. Using this model as a one-step-ahead predictor, the prediction error variance over a suitable averaging time is determined. It is expected that this error variance will be larger during fault conditions than during normal operating conditions. The mean square prediction error (MSPE) averaged over $M$ samples is given as

$$MSPE = \frac{1}{M}\sum_{j=1}^{M}\left|e[j]\right|^2$$

$$MSPE = \frac{1}{M}\sum_{j=1}^{M}\left|x[j] - \tilde{x}[j]\right|^2,$$

where $\{x[j]: j = 1,2...,N\}$ is the actual signal value, $\{\tilde{x}[j]: j = 1,2,...,N\}$ is the predicted value, $N$ is the total number of samples of the signal and the error term at any value $j$ is given by the difference between the observed and predicted values.

The predicted values are obtained by using the estimated model parameters $\{\tilde{a}[k]: k = 1,2,...,p\}$ in the following expression:

$$\tilde{x}[j] = -\sum_{k=1}^{p}\tilde{a}[k]x[j-k].$$

Note that the mean square of the prediction error is dependent on the model order, $p$, and the number of samples $M$. This latter parameter effectively determining the total time taken to reach a decision of whether or not an anomalous event has occurred .

### A.5.2    Distance-in-parameter-space

A more direct method, developed by Simon Rofe, is to use a feature which incorporates the prediction error variance as follows:

$$\textbf{Feature} = \left\| \frac{a_{ref}}{\sigma_{ref}} - \frac{a_{test}}{\sigma_{test}} \right\|,$$

where $a_{ref}$ and $a_{test}$ are vectors of the estimated reference and test parameters, and $\sigma_{ref}$ and $\sigma_{test}$ are the standard deviations of the reference and test signals respectively. The reference is obtained from signals at the start of the file which are known to be 'normal' and the test is obtained from signals at all subsequent times. The basic assumption used in this method is that the model structure will be different under fault conditions when compared to normal operation. The standard deviation is used in this feature to normalise the dimensions of the feature.

### A.6    RESULTS OF FEATURE EXTRACTION STRATEGIES

The results of the two fault detection strategies for each of the five signal-to-noise ratios used in the tests follow. In each case the estimates of the parameters used in feature extraction were calculated over 20,000 samples of data collected over less than 15 ms.

Figures A.7 to A.11 show the mean-square prediction error for signal-to-noise ratios from −6dB to −24dB. The results of the distance-in-parameter-space method are shown in Figures A.12 to A.16 for the same signal-to-noise ratios.

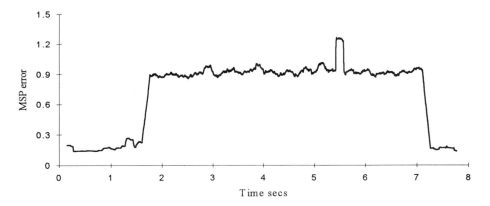

**Figure A.7** Mean square prediction error versus time from start of file at −6dB S/N

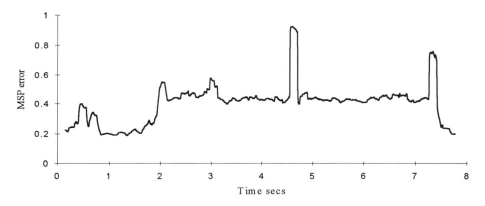

**Figure A.8** Mean square prediction error versus time from start of file at −12dB S/N

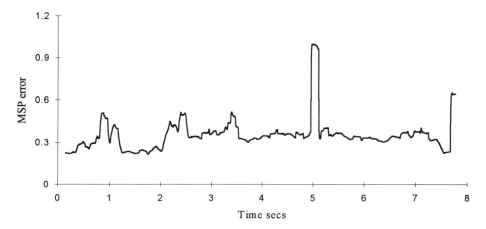

**Figure A.9** Mean square prediction error versus time from start of file at −16dB S/N

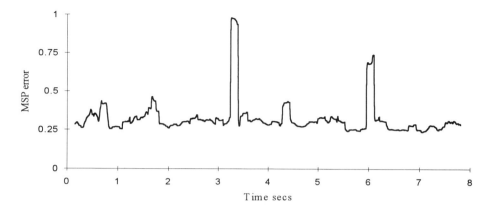

**Figure A.10** Mean square prediction error versus time at −20dB S/N

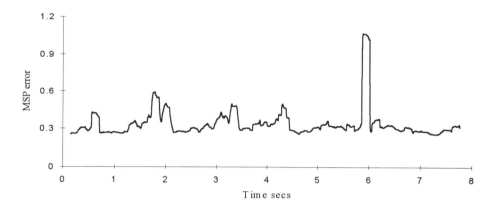

**Figure A.11** Mean square prediction error versus time at −24dB S/N

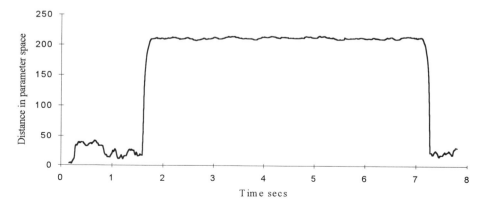

**Figure A.12** Distance-in-parameter-space versus time at −6dB S/N

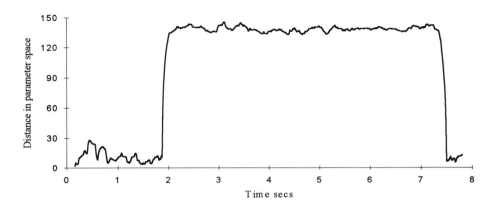

**Figure A.13** Distance-in-parameter-space versus time at −12dB S/N

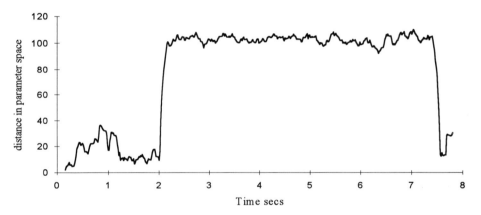

**Figure A.14** Distance-in-parameter-space versus time at −16dB S/N

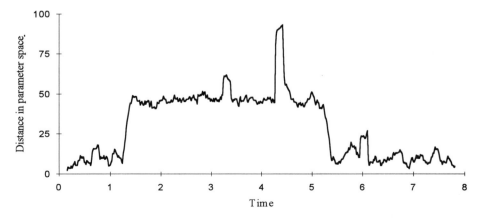

**Figure A.15** Distance-in-parameter-space versus time at −20dB S/N

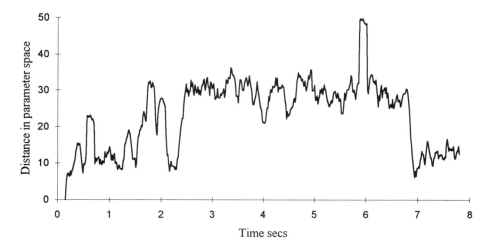

**Figure A.16** Distance-in-parameter-space versus time at −24dB S/N

## A.7 DISCUSSION OF RESULTS OF FEATURE EXTRACTION

Comparisons between the relevant Figures for each of the five signal-to-noise ratios show that the distance-in-parameter space method is markedly superior to the mean-square prediction error method and also to the likelihood ratio test outlined in the full paper but omitted here for the sake of brevity.

It is possible to use the distance-in-parameter space method to estimate the onset and duration of fault conditions at all signal-to-noise ratios used. In contrast, the mean-square prediction error shows a clear distinction between normal and fault conditions only at the highest signal-to-noise ratio.

The estimates for the start of the leak signal and its duration are shown in the following Table A.1, which also shows the signal-to-noise ratios inferred from the relative outputs of the feature extraction strategy by assigning the highest signal-to-noise ratio to the clearest indication of the onset of an anomaly.

**Table A.1** Showing estimates obtained using distance-in-parameter-space method.

| Test set | Time of start (Seconds) | Time of end (Seconds) | Signal-to-noise ratio |
|----------|------------------------|----------------------|----------------------|
| 0 | 1.87 | 7.50 | −12dB |
| 1 | 1.23 | 5.39 | −20dB |
| 2 | 2.01 | 7.56 | −16dB |
| 3 | 1.60 | 7.26 | −6dB |
| 4 | 2.3 | 6.94 | −24db |

**Table A.2** Actual values of test data supplied

| Test set | Time of start (Seconds) | Time of end (Seconds) | Signal-to-noise ratio |
|----------|-------------------------|-----------------------|-----------------------|
| 0 | 1.875 | 7.34 | −12dB |
| 1 | 1.23 | 5.23 | −20dB |
| 2 | 2.01 | 7.40 | −16dB |
| 3 | 1.61 | 7.11 | −6dB |
| 4 | 2.29 | 6.79 | −24db |

At the research coordination meeting held in India in November 1994, participants were provided with details of the actual test data values as generated by the producers of the test data sets. These values are given in Table A.2 and were used to compare the effectiveness of the various signal processing strategies employed in the coordinated research programme.

The distance-in-parameter space method correctly identifies the start time of each fault condition and is reasonably close to identifying the end. The inferred ranking of the signal-to-noise ratio is absolutely correct.

## A.8     TIME DELAY AND LOCATION ESTIMATION

The secondary major task was to attempt to locate the spatial coordinates of the leak in the test rig. It was decided to examine the application of cross-correlation methods to this task and use the normalised cross-correlation function defined in Chapter 2 as a basis for the examination. This function is defined as

$$\phi_{xy}(\lambda) = \frac{R_{xy}(\lambda)}{\sqrt{R_{xx}(0)R_{yy}(0)}}.$$

It was anticipated that the medium through which the acoustic energy was propagated would be dispersive, and therefore the envelope and the mean square of the signals were used in these estimates. Both these measures gave reasonably high correlation coefficients in the range 0.4 to 0.7, compared to the use of the raw signal which gave coefficients in the range −0.28 to 0.078. It was concluded, therefore, that the medium was highly dispersive. Typical correllogrammes are shown in Figures A.17, A.18 and A.19.

Attempts to use the measured delay obtained from the cross-correlation function were not successful due to the extremely complex geometry of the installation and the existence of multiple paths along which the acoustic energy travelled. Perhaps a more extensive array of sensors would produce better results.

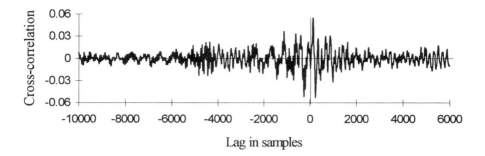

**Figure A.17** Cross-correlation function, raw data, sensors 1 and 2

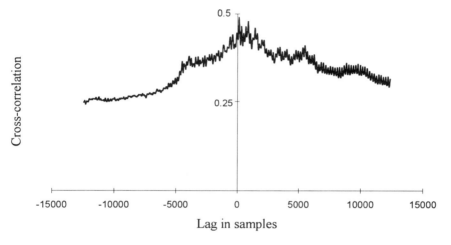

**Figure A.18** Cross-correlation function, envelopes of the signals, sensors 1 and 2

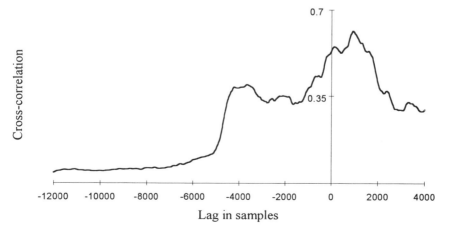

**Figure A.19** Cross-correlation function, mean square of the signals, sensors 1 and 2

## A.9      CONCLUSIONS

- The detection of the onset of a leak in a steam generator unit of a liquid metal cooled fast reactor is possible using Autoregressive modelling approaches, the distance-in-parameter space method proving to be the superior method.
- Spectral composition of the signals from each of the four sensors used is significantly different from the others, indicating that a generic method of fault detection is the most appropriate one.
- Correlation analysis of multi-sensor data has shown that envelope detection or use of the mean-square produces correlation coefficients very much greater than those obtained using the raw signal. Therefore the medium is dispersive.
- Time-delay estimation, and hence location of the leak, is difficult to interpret in the particular physical environment encountered. The problem created by the presence of a dispersive media is compounded by the existence of multiple paths for the transmission of acoustic energy. It is possible for the acoustic noise source to be distributed over considerable space.

**REFERENCES**

Akaike, H., 1970, A new look at statistical model identification, *IEEE Transactions on Automatic Control*, AC-19, 716-723.

Kay, S. M. & Marple, S. L. 1981, Spectrum analysis - a modern perspective, *Proceedings of the IEEE*, 69 (11), 1380-1419.

Kay, S. M., 1988, *Modern Spectral Estimation: Theory and Applications*, Prentice Hall.

Marple, S. L., 1987, *Digital Spectral Analysis: with applications*, Prentice Hall.

Rissanen, J., 1983, A universal prior for integers and estimation by shortest data description, *Annals of Statistics* 11 (2), 416-431.

Rofe, S. E. F. & Ledwidge, T. J., 1994, *IAEA Extended Co-ordinated Research Programme on Acoustic Signal Processing for the Detection of Boiling or Sodium/Water Reaction in LMFBR*, Report to the research co-ordination meeting, Kalpakkam, India, 1-3 November.

# Case Study B

## DETECTION OF BOILING IN A NUCLEAR REACTOR CHANNEL

This case study is an edited version of a paper by G. J. Cybula, R. W. Harris and T. J. Ledwidge

### B.1    CONTEXT

In a water-cooled nuclear reactor the limit of operation is set by the physics of the heating/boiling of the liquid coolant. When the power is low and water flows at its design value over the fuel elements, heat is transferred by convection, the temperature difference between the surface of the fuel element and the surrounding fluid being small. If the power, and hence the heat flux, is raised then ultimately a superheated layer forms around the fuel element and small vapour bubbles form at nucleating sites on the heated surface. This stage is known as the onset of nucleate boiling. Further increases in heat flux result in the generation of an ever-increasing vapour bubble population, culminating in coalescence of the vapour bubbles which produces a vapour blanket around the fuel element. At this point, when the predominant heat transfer is radiation, the temperature of the surface of the fuel element increases substantially and, for all practical purposes, sets the upper limit of operation. The heat flux at which this limit occurs is the critical heat flux (CHF[1]).

Ideally the measurement of the CHF for a given nuclear fuel channel would be obtained from experiments in an electrically heated test rig operating at the same temperature, pressure and range of heat flux as that in the reactor. This approach makes severe demands on the electrical power supply and also on the mechanical structure of the test rig. Usually scale models are used to estimate the value of CHF when the geometry is simple and where the scaling may encompass the physical size and the form of coolant.

The use of Freon to model the CHF behaviour of high-pressure water in tubes is accurate, simple and inexpensive (Groeneveld 1970). Freon has the same vapour/liquid ratio as water but a lower latent heat of vaporisation, resulting in the test section power of a Freon system being about 6% of the

---

1   Other terms used to denote CHF are the burnout heat flux and the dryout heat flux.

equivalent water system and the operating temperature being 20% of that in the equivalent water system. With these considerably reduced conditions, test loops can be constructed to include a transparent section permitting flow visualisation studies to be conducted.

Often other measurement techniques are used in conjunction with flow visualisation in an attempt to use their findings to monitor actual nuclear fuel channels when flow visualisation is clearly not possible.

## B.2     MEASUREMENT TECHNIQUES

The measurement technique used to detect the onset of boiling in a channel depends on the type of liquid, the temperature, the pressure and the physical structure at the point of interest. If the liquid is electrically conductive, an integrating conductivity probe can be useful. If the fluid is an insulator, then a capacitance void gauge may be employed. Both these methods require penetration of the wall of the channel and are not suitable in a nuclear reactor. Void detectors based on the attenuation of gamma rays do not require any penetration of the wall of the channel but are not considered practical for in-pile application.

The method described here relies on the detection of the acoustic noise produced by the growth and collapse of vapour bubbles during the nucleate boiling phase of the heat transfer process. This method has considerable advantages over the other methods mentioned, as it can be applied in a reactor system to detect the onset of nucleate boiling and, in principle, locate the spatial coordinates of the site of the boiling. This method was developed as part of an investigation into heat transfer phenomena in annular fuel elements and uses conventional accelerometers clamped to the outside of the test rig to detect the onset of nucleate boiling.

The experimental arrangement shown in Figure B.1 utilised a transparent section in a test rig in which Freon 12 was pumped around a closed loop under controlled conditions. The transparent section contained a stainless steel tube conected to a variable power supply. At any given flow condition the power could be raised to a level at which boiling commenced within the test section. The acoustic noise produced by the nucleate boiling process was detected by two accelerometers; one at each end of the test section as shown in Figure B.1. The accelerometers were standard commercial ones, each with a useful bandwidth of 40 kHz. The outputs from the two accelerometers were recorded as analogue signals on an instrumentation tape recorder for later analysis, as discussed in sections B.4 and B.5.

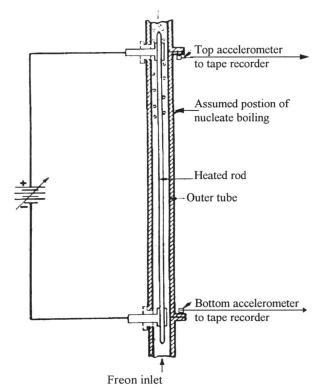

Top accelerometer
to tape recorder

Assumed postion of
nucleate boiling

Heated rod

Outer tube

Bottom accelerometer
to tape recorder

Freon inlet

**Figure B.1** Test section simulating a nuclear fuel rod

## B.3  DETRIMENTAL EFFECTS ON THE CROSS-CORRELATION

Although the detection of the onset of nucleate boiling is a straightforward exercise the location of spatial coordinates is not. A natural tendency is to use cross-correlation techniques to measure the time difference between the arrival of acoustic signals and thus estimate the position of the boiling using this data and the measured velocity of propagation of the acoustic energy. There are two quite different phenomena which have very significant effects on the cross-correlation function and need to be accounted for by some form of pre-processing. These are the frequency deviations caused by differences in the rigidity of the clamping of the accelerometers and the dispersive nature of the medium through which the acoustic energy must pass.

### B.3.1  The Effect of Frequency Deviation on Cross-correlation

The flanges to which the vibration sensors are attached are not necessarily identical, which means that the natural modes of vibration of flanges may be different, resulting in the frequencies recorded by the two sensors being slightly different.

To illustrate this point, consider the case in which one sensor responds at a frequency $\omega$ and the other sensor at a frequency $\omega + \delta\omega$ delayed by an amount $\tau$, the transport delay between the two signals. In this case the cross-correlation between these two signals, when averaged over a time $T$, is given as

$$R_{xy}(\lambda) = \lim_{T \to \infty} \frac{1}{2T} \int_{-T}^{+T} A \sin(\omega[t - \lambda]) B \sin([\omega + \delta\omega][t - \tau]),$$

which can be approximated as

$$R_{xy}(\lambda) = \frac{AB}{8} \frac{\sin(\delta\omega T / 2)}{\delta\omega T / 2} \cos(\delta\omega T / 2) + [(\omega + \delta\omega)\lambda - \omega\tau]).$$

When the two sensors respond at exactly the same frequency, $\delta\omega = 0$, and thus the correlation function is a cosine wave. The intrinsic delay between the two sensors can then be determined by the location of the first maximum of this cosine function.

The effect of various frequency deviations on the cross-correlation function is shown in the following Table for a range of averaging times $T$.

Note that for even small differences in the frequencies detected by the sensors, the correlation function rapidly approaches zero as the averaging time is increased. Usually long averaging times are necessary to average out the effects of uncorrelated and hence unwanted noise sources.

A slightly more complicated situation further illustrates this phenomena. Consider a wave packet made up of five frequencies with the values and weightings defined in Table B.2.

Figure B.2 shows the effect of correlating the wave packet with a delayed version of itself with a 10 Hz frequency deviation applied to each of the constituent frequencies. 10 Hz is equivalent to an average deviation of only 0.2%. Larger deviations again result in the correlation function converging rapidly to zero.

**TableB.1** Illustration of averaging time for various frequency deviations

| Frequency deviation $\delta\omega$ r/sec | Value of the cross-correlation function for various averaging times $T$ in seconds | | |
|---|---|---|---|
| | $T = 1$ | $T = 10$ | $T = 100$ |
| 0 | 1 | 1 | 1 |
| 0.1 | 0.99 | 0.95 | -0.19 |
| 1.0 | 0.95 | -0.19 | -0.005 |
| 10 | -0.19 | -0.005 | -0.0009 |

**Table B.2** Composition and weighting of a wave packet

| Number ($N$) | Frequency (Hz) | Weighting factor |
|:---:|:---:|:---:|
| 1 | 3400 | 0.5 |
| 2 | 4200 | 0.8 |
| 3 | 5200 | 1.0 |
| 4 | 5800 | 0.8 |
| 5 | 6600 | 0.5 |

## B.3.2  The Effect of Dispersion on Cross-correlation

Dispersion occurs when the velocity of propagation is a function of frequency and in some cases is also a function of the geometry if the medium is bounded. For example, consider the case of longitudinal elastic waves being propagated along a solid bar of circular cross -section. In this case the velocity of propagation is given as

$$C = C_0[1 - \sigma\pi(a/\Lambda)^2],$$

where

$C_0 = \sqrt{\dfrac{E}{\rho}}$  the velocity in an unbounded medium

$E =$  Young's modulus of elasticity

$\rho =$  Density of the material of the rod of radius a

$\sigma =$  Poisson's ratio

$\Lambda =$  Wavelength of the propagation.

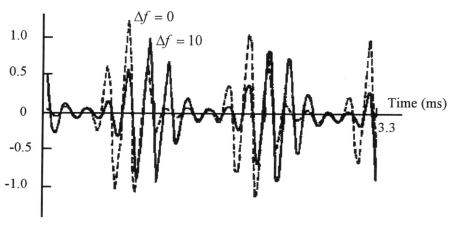

**Figure B.2**  Comparison of cross-correlograms for a frequency deviation of 10 Hz with one for zero deviation

This variation of the velocity with wavelength (or frequency) means that the transit time between the two sensors will be different for each part of the spectrum of the propagation and thus the cross-correlation function will be 'smeared' out. In extreme cases this smearing out will render the cross-correlation function completely useless.

To illustrate this point consider the wave packet in Table B.2. In addition, the transit time chosen to be related to the number $N$ of the wave packet by using the following algorithm which closely models the delay found in the practical exercise reported later.

$$\text{Delay time} = 8.25 \times 10^{-4} [1-(N-3)\Delta]$$

The results are summarised in Table B.3. Note once more that the effect of dispersion can be significant if the transit time deviation factor is greater than about 5% , for example.

## B.4    SIGNAL PROCESSING STRATEGY

The previous two sections showed that both frequency deviation and dispersion cause degradation in the computed value of the cross-correlation function, especially if long averaging times are used. However, in practice long averaging times are usually required to minimise the contribution due to uncorrelated noises in the two signals.

In this experiment, to overcome both these effects, two forms of pre-processing were employed. The first signal processing technique is one that is widely used and consists of filtering a narrow band of frequencies in which the contribution from the boiling process is greater than that from the background noise.

The second signal processing strategy relies on the recognition that the boiling process generates pulses of acoustic energy occurring at a random epoch. The envelope of the acoustic signal is propagated with a group velocity whilst the individual frequencies which comprise the fine detail of the signal each have their own phase velocity. The group velocity is less sensitive to the effects of dispersion than the phase velocity, because it is essentially a long wavelength phenomena. Similarly the envelope detection method is relatively

**Table 3** Illustration of the effects of transit time deviation

| Transit time deviation factor $\Delta$ | Normalised cross-correlation function |
|---|---|
| 0.100 | 0.66 |
| 0.010 | 0.97 |
| 0.001 | 1.00 |

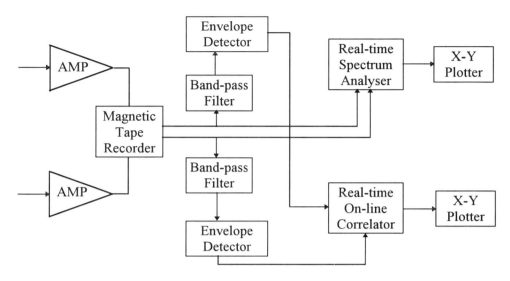

**Figure B 3** Instrumentation arrangement and signal flow paths

unaffected by any variation of the resonant modes of the flanges to which the vibration sensors are attached.

The instrumentation arrangement used to implement this signal processing strategy is shown in Figure B.3.

## B.5    EXPERIMENTAL RESULTS

The power to the test section was raised slowly until vapour bubbles were clearly visible in the test section. Simultaneously, the power spectral density recorded at both sensors changed significantly. (In many cases the change in acoustic signal was detected before the vapour bubbles were visible.) A comparison between the power spectral density obtained while boiling was in progress to that when the power to the section was off, gave the basis for using the band pass filter in further analysis.

A typical comparison between the spectra due to boiling and non-boiling is shown in Figure B.4. Note that a frequency of 17.4 kHz was chosen as the centre of the bandpass filter in preference to a frequency around 14 kHz, because the latter did not appear as stable as the former over a number of repeated trials.

The location of the point of the onset of nucleate boiling required an estimate of the time delay between the signals at each of the two sensors which could be determined from the data given in Figure B.5. The first and fourth

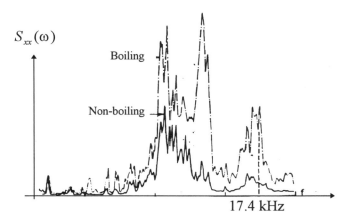

**Figure B.4** Typical power spectral densities of boiling and non-boiling regimes

traces in the Figure show the raw signals received from the two accelerometers at the instant nucleate boiling started. The high frequency bursts which can be seen in each trace were due to a vapour bubble growing or collapsing. For each burst in the top trace there is a corresponding burst in the fourth trace, with an apparent, but difficult to measure, delay between the two signals.

Attempts to use the cross-correlation method between the two raw signals to measure the time delay between the first and fourth traces failed to produce any correlation, due to the effects discussed in section B.3.

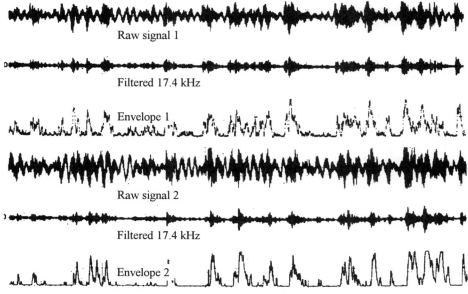

**Figure B.5** Signals from the top and bottom accelerometers at various stages of processing

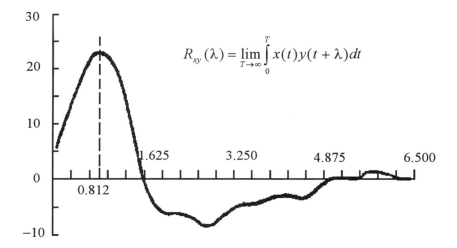

**Figure B.6** Typical cross-correlation function of the envelopes of the filtered signals taken while nucleate boiling was in progress

The two raw signals were then passed through a narrow bandpass filter set at 17.4 kHz, then demodulated by rectification and smoothing before being subject to a cross-correlation routine. The second and fifth traces show the outputs of the narrow bandpass filters and the third and last traces show the demodulated or envelopes of the events. The apparent time delay between the two sensors is seen more clearly in the latter case and made explicit by the cross-correlation function shown in a typical result in Figure B.6.

The information in Figure B.6 was crucial to the determination of the location of the noise source as it gave the time delay of 0.81 ms between the arrival of the signal at the top and bottom sensors.

The velocity at which the acoustic energy propagates was estimated by holding the conditions in the test section just below those required to cause boiling, striking the pipe at a point just below the lower sensor and measuring the resulting time delay between the filtered envelope detected signals. This velocity was found to be 1496 m/s (probably the propagation was a flexural wave).

Using the estimated velocity of propagation, the location of the site of the nucleate boiling was estimated as 0.61 m below the position of the upper sensor.

## B.6    CONCLUSIONS

- Cross-correlation of the envelope of filtered signals can yield useful information about time delays in the system in some circumstances where the correlation of the raw signal can not.

- The method of acoustic detection is non-invasive and thus has significant advantages over other techniques that require penetrations into the test channel.
- The method of cross-correlation of the envelope of filtered signals can be applied to a wide variety of industrial problems that may otherwise be intractable.

## REFERENCES

Cybula, G. J., Harris, R. W., & Ledwidge, T. J. 1974, Location of a noise source by noise analysis techniques *Proceedings of the IREE*, 35, No 10, October, 310-316.

Groenveld, D. C., 1970, Similarity of water and Freon dryout for uniformly heated tubes, *ASME Paper 70-HT-27*.

# Case Study C

## GENERIC FAULT DETECTION STRATEGIES IN LIQUID METAL FAST BREEDER REACTORS

This case study is an edited version of a paper by S. E. F. Rofe and T. J. Ledwidge presented to a research coordination meeting of the International Atomic Energy Agency held in Obninsk, Russia, 11-14 July 1995.

### C.1    CONTEXT

The detection of faults in liquid metal cooled fast reactors is in the same class as the detection of faults in any industrial process, but differentiated from them by the high power density associated with fast reactors. If left unchecked, a fault in a fast reactor could lead to considerable damage to the plant and become a threat to human life. Therefore the early detection of incipient mal-function is of paramount importance in the operation of all fast reactors.

The two components of the plant which are of significant concern are the reactor core and the heat exchangers. Protection of the core is problematic because of the high power density, and the heat exchanger presents special threats because of the possible interaction between water and sodium in the event of a leak in any of the tubes in the heat exchanger. This is because the steam generators in the prototype fast reactor (PFR) at Dounreay in the north of Scotland, are of the shell and 'U' tube design with water or steam flowing through the tubes and sodium flowing over the outside of the tubes. There are three essentially identical secondary circuits, each comprising an evaporator, a superheater and a reheater.

It may be possible, by extensive in-plant testing, to characterise the features of defined fault conditions in a particular plant and design a fault detection strategy around the data obtained in this way. Srinivasan and Singh (1990) demonstrated a method showing impressive discrimination by using a selected set of individual frequencies from a spectral estimate in a covariance matrix. However, it may not always be possible to transfer the results of these in-plant tests to another situation with a high degree of confidence unless the plants are identical in design.

In the work under review, generic fault detection strategies were examined in order to determine which ones are applicable to the detection of incipient malfunction in a liquid metal cooled fast reactor. Ideally a generic strategy for fault detection should not make any assumptions about the nature of the fault conditions. Although this means the strategy is an anomalous event detector, it has the advantage that faults not previously considered may be identified, otherwise these faults may pass unnoticed.

Data used to test a variety of generic fault detection strategies were provided as part of the IAEA coordinated research programme on acoustic signal processing for the detection of boiling or sodium/water reaction in a liquid metal cooled fast breeder reactor.

## C.2    THE TEST DATA

Test data were supplied in 1993 and 1994 in the form of signals from the superheater in the prototype fast reactor (PFR) in Scotland, mixed with signals from an experimental rig in Bernsberg in Germany. The test data supplied in 1995 were obtained from the PFR during de-commissioning and comprised background signals from both the evaporator and superheater in secondary circuit number three operating at full power, and a separate set of recordings made while injecting argon or hydrogen or water into the evaporator during the PFR shut down.

**The 1993 data** (supplied by the United Kingdom Atomic Energy Authority). consisted of a calibration signal followed by 12 recordings The first was the background noise from the PFR superheater number two and the second was a 5 s sample of the sodium-to-water injection at a rate of 1.8 g/s. The next four records consisted of the signals from the background noise mixed with the noise from an injection starting at an arbitrary time with a signal-to-noise ratio set at one of the values from a maximum of −12 dB to a minimum of −24 dB. This arrangement of test data was repeated with PFR superheater number three using a 5 s sample of a sodium-to-water injection at a rate of 3.8 g/s.

**The 1994 data** comprised multichannel recordings prepared by CEA, Cadarache, France. Using the original 1993 data, four channels of test data were generated by mixing the signals from a leak detected by four sensors attached to waveguides on the test vessel in Germany with signals from four sensors attached to waveguides on the PFR superheater number two. Five signal-to-noise ratios from −6 dB to −24 dB were used to prepare the test data from analogue signals sampled at 131 kHz. A more detailed explanation of the 1994 data is given in section A.2 in Case Study A.

Although similar waveguides and sensors were used for both the 1993 and 1994 benchmark tests, an uncertainty remained on the general applicability of the findings because the signals from the simulated fault were obtained in a different plant to that from which the background signals were obtained, and the usefulness of these trials rested on identifying signal processing strategies that showed potential for use in real industrial situations.

**The 1995 test data** provided by the Atomic Energy Authority, Reactor Technology Department, Dounreay, Caithness, Scotland were obtained using precisely the same waveguides and sensors on the same item of plant under normal operation and test conditions. Thus the uncertainly associated with the use of different plant for the generation of background and fault signals was removed.

The data sets used in this work were derived from recordings of the output of four accelerometers attached to waveguides on the evaporator in secondary circuit number three during the injection of hydrogen, argon and water at various rates of flow. One of these injections, designated as injection number five, was selected as the test sample for comparison of the signal processing strategies in this case study. The accelerometers were Endevco type 7704A-17, having a typical sensitivity of 17 pC/g and a resonant frequency of 45 kHz. The signals from each of the four accelerometers were recorded in FM mode at a tape speed of 120 inches per second, giving a nominal frequency range of 0 to 80 kHz. These signals were subsequently digitised, two at a time, by using a LSI DSP32-C digital signal processing card in a Dell 486D/33 PC computer. In order to overcome the limitations of using only two signals at a time and to facilitate the use of multichannel methods of fault detection and location, the signals were digitised synchronously in pairs. The first pair were designated sensors A and B, the second pair C and B, and the third pair D and B. In this way the original synchronised recordings of all four sensors could be reconstructed.

## C.3      GENERAL SIGNAL PROCESSING STRATEGY

The development of a generic signal processing strategy is based on two assumptions. It is assumed that the background noise is spatially distributed around the plant giving rise to a signal at a sensor which is sampled appropriately to generate a series $x[j]$. However, the noise associated with a fault is assumed to be localised and thus is not necessarily correlated with the background noise. The noise associated with the fault gives rise to a signal at the sensor which is sampled to generate a series $y[j]$ which is regarded as the target signal.

In general, the sampled signal from the sensor may be written as

$$z[j] = x[j] + \delta \times y[j],$$

$$\text{where } \begin{array}{l} \delta = 0 \text{ corresponds to no fault and} \\ \delta = 1 \text{ corresponds to fault conditions.} \end{array}$$

Although this model is simple, it has significant implications for generic fault detection strategies because the fault acts in an additive sense and will produce a change in the structure of the signal model which can be observed from the sensor. Because no other *a priori* information is required, additive changes in descriptions of $z[j]$ in the time, frequency or probability domains can be deduced.

## C.4       SPECIFIC SIGNAL PROCESSING METHODS

Five different signal processing methods are reviewed in this section, i.e.
Spectral distance measures
- *direct spectral distance*
- *normal spectral distance*

Wavelet analysis
- *mean square of the wavelet decomposition signals*
- *direct spectral distance of wavelet decomposition signals*
- *normal spectral distance of wavelet decomposition signals.*

## C.4.1      Spectral Distance Measures

A direct way to attempt detection of the onset of fault conditions is to compare the spectral components of a reference block of data to that of test blocks throughout the signal record. This method is generic in that it does not use any *a priori* information. However, it may incorporate significant redundancy and loss of resolution by including all parts of the frequency spectrum in the analysis.

Two variations are possible using this approach, and for convenience are called the *direct spectral distance measure* and the *normal spectral distance measure*.

### C.4.1.1    *Direct Spectral Distance Measure*

This method compares the frequency components of the fast Fourier transform (FFT) of a reference set of data to that of test blocks by forming each of the FFTs into a vector of size P.

In this method the original signal is divided into a number of blocks of data each containing M samples. Depending on the size of M, it may be necessary to further subdivide to facilitate the use of available FFT algorithms and take the average over these subdivisions. The resulting vector of the spectral components at the ith level of decomposition is written as $X_i(f)$ and the distance measure for the detail signal at this level is given as:

$$D_i(j) = \sum_{k=1}^{M} \left[ |X_{i,ref}[k]| - |X_{i,test}[k]| \right] \quad j = 1,2,3,\ldots,K.$$

In this case 20,000 samples of the original signal were selected for analysis. These samples were divided into 78 contiguous blocks, each containing 256 samples, thereby using 19,968 of the available data set. The choice of 256 samples in each block facilitated the use of a standard FFT algorithm to generate the FFT of each of the 78 blocks. The FFT of the individual blocks is

$$X_i[f] = FFT\{x_i[j]: i = 1,2,\ldots,78\}$$

$$\text{for } j = 1,2,\ldots,256$$

$$\text{and } f = 1,2,\ldots,128.$$

Note that the range for the frequency $f$ is only taken to 128, corresponding to the folding frequency. The 78 FFTs were then averaged to produce a vector with elements at each of the 128 frequencies. This process was implemented for both the reference and test signals to yield the following vectors

$$\mathbf{X}_{ref} = \left[ X_{ref}[1],\ldots,X_{ref}[128] \right],$$

$$\mathbf{X}_{test} = \left[ X_{test}[1],\ldots,X_{test}[128] \right].$$

The feature is then the Euclidean distance between the reference and test vectors; that is,

$$\left\| \mathbf{X}_{ref} - \mathbf{X}_{test} \right\|.$$

## C.4.1.2  Normal Spectral Distance Measure

One view of the spectral distance measure first proposed by Black et al. (1993) is that differences in the direction of the reference vector only reflect an increase in the volume of the background signal and thus may not be indicative of the on-set of fault conditions. He proposed that a distance measure based on deviations normal to the reference vector would indicate a change in shape of the power spectral density function and be more likely to signal the onset of fault conditions.

If the reference and test vectors are denoted by **r** and **t** respectively, and a vector orthogonal to **r** in the direction of **t** is **q**, then it can be shown that,

$$x = t\text{-}q\text{-}r$$

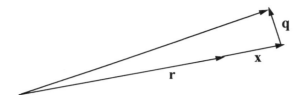

**Figure C.1** Illustration of normal spectral distance measure

where **x** is a vector in the direction of **r**. A two dimensional example is shown in Figure C.1.

The feature proposed by Black was the square of the magnitude of the normal and is

$$|\mathbf{q}|^2 = \mathbf{t}.\mathbf{t} - \left[\frac{\mathbf{r}.\mathbf{t}}{|\mathbf{r}|}\right]^2,$$

representing the magnitude of deviation of the test vector in a direction normal, to the reference vector. This expression can be written with the same convention as used earlier; i.e.

$$\text{feature} = \mathbf{X}_{test}.\mathbf{X}_{test} - \left[\frac{\mathbf{X}_{ref}.\mathbf{X}_{test}}{|\mathbf{X}_{ref}|}\right].$$

### C.4.2   Wavelet Analysis

The wavelet transform decomposes a signal space into a series of subspaces, usually by filtering. This concept is illustrated in Figure C.2 in which a signal $x[j]$ in the space $V_0$ is decomposed into an *approximation* signal in $V_1$ together with some added *detail signal in* $W_1$. The space $V_1$ contains all that part of the

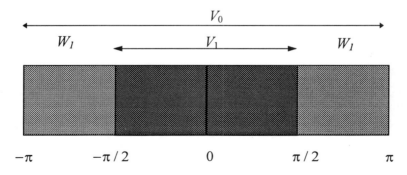

**Figure C.2** Division of space $V_0$ into subspaces $V_1$ and $W_1$

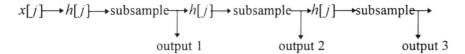

$$x[j] \longrightarrow h[j] \longrightarrow \text{subsample} \xrightarrow{} h[j] \longrightarrow \text{subsample} \xrightarrow{} h[j] \longrightarrow \text{subsample} \longrightarrow$$

output 1                    output 2                    output 3

**Figure C.3** Part of the discrete wavelet transform filter when subsampling by two

signal limited to the range $(-\pi/2, \pi/2)$ and may be regarded as the output of an ideal low-pass filter. $W_1$ contains the rest of the signal, which from a signal processing perspective may be taken as the output of an ideal bandpass filter. The frequency scale $(-\pi, \pi)$ may be considered the folding frequency on a normalised frequency scale.

In general, the signal space $V_0$ may be decomposed (divided) into a number of related subspaces by sub-sampling. In this work every alternate sample was used. Subsampling by two was used for convenience here, although it is possible to select any sample set. The major advantage of subsampling by taking every second sample is that the wavelets generated are orthonormal and hence the different decomposition signals are (ideally) independent.

Subsampling is of fundamental importance to the discrete wavelet transform since it allows the same filter coefficients to be used at each level of decomposition rather than scale the filter to match the bandwidth of concern. In the current work the wavelet transform was implemented using the filter coefficients $h[j]$ defined by Daubechies (1988) using the computational scheme shown in Figure C.3 where the *detail* outputs are equivalent to constant percentage bandpass filters. Outputs corresponding to the *approximation* of the input are not necessary in this case. The actual bandwidth at each level of decomposition is shown in Table C.1 where the bandwidth at any particular level is one-half that of the bandwidth at the level above, except in the final bandwidth, which is rounded off to a lower bound of zero. The whole of the usable frequency band is covered by the use of ten levels of decomposition and thus the method can be regarded as generic.

**Table C.1** Bandwidth for each level of decomposition

| Level | Frequency Range, kHz |
|---|---|
| 0 | 40-80 |
| 1 | 20-40 |
| 2 | 10-20 |
| 3 | 5-10 |
| 4 | 2.5-5 |
| 5 | 1.25-2.5 |
| 6 | 0.625-1.25 |
| 7 | 0.3125-0.625 |
| 8 | 0.15625-0.3125 |
| 9 | 0-0.15625 |

The mean squares of the outputs of the equivalent constant percentage band-pass filters form a feature that may be used to detect the onset of an anomalous event. In practice, attention would be focussed on those wavelets in which the energy associated with the anomalous event predominantly lies. Wavelets can be used by themselves or in combination with either of the two spectral distance measures, giving three possible combinations in all. When using wavelets by themselves, the calculation of the mean square or the variance is estimated from the samples within each level of decomposition. In the case under review, decomposition level zero contained 8192 samples, level 1 contained 4096 samples, level 2 2048 samples and, dividing by two each time, finally 32 samples in each of levels 8 and 9 remained. The samples in each of the decomposition levels were used as the inputs to the same FFT algorithm used previously to generate reference and test vectors in precisely the same way as was implemented in the calculation of the direct and normal spectral distances explained in sections C.4.1.1 and C.4.1.2.

The three variations associated with wavelets, i.e. mean square (or variance), direct spectral distance of wavelet decomposition signals, and normal spectral distance of wavelet decomposition signals were thus estimated from the same set of sampled data obtained by subsampling the original discrete data by two.

## C.5      RESULTS OF THE FEATURE EXTRACTION STRATEGIES

The results presented in graphical form in the Figures that follow relate to trials of the feature extraction strategies with data from the 1994 and 1995 benchmark tests. In order to compare the various signal processing strategies under conditions as closely similar to each other as possible, the 1995 data supplied were modified by mixing the background noise from the evaporator with the noise from a typical injection. The results reported here for the modified 1995 data are for injection number 5 using sensors A and B only to limit the volume of the data processing to manageable proportions. The methods investigated may be directly extended to the other injections and other sensors, without any need for further modification.

Figure C.4 shows the results of computing the FFTs of the noise signal of the evaporator background and that of injection number 5 which had been recorded separately. The noise from the injection appears to excite the same set of harmonics as are excited by the evaporator noise but with a significant difference in the spectra below 1 kHz. Figures C.5 and C.6 show the results of analysis implemented on the signals formed by adding together the signal from the injection to that of the background noise. Figure C.5 shows the effects of applying direct and normal spectral distance measures and the distance-in-parameter-space method. (This method is discussed in detail in Case Study A.)

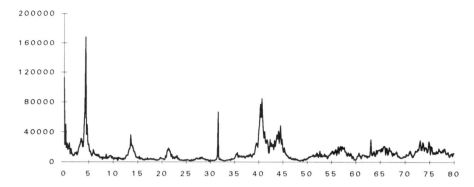

(i) Spectrum of evaporator background noise to 80 kHz

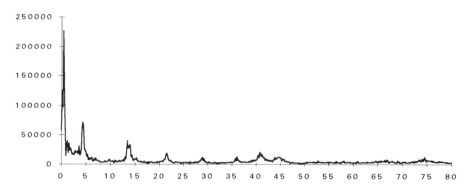

(ii) Spectrum of signal of noise from injection 5 to 80kHz

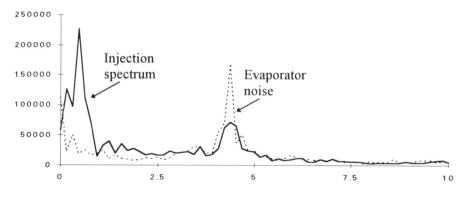

(iii) Comparison of signal spectra from injection 5 noise to evaporator noise

**Figure C.4** Frequency spectra of noise from evaporator and injection 5
using 1995 data

Use of the three variations of wavelet analysis is shown in Figure C.6 for wavelet decomposition level 6. This level of decomposition corresponds to a bandpass of 0.625 kHz to 1.25 kHz and was the decomposition level showing the clearest indication of the onset of the injection. This effect was expected because the energy associated with the injection, as indicated in the spectrum shown in Figure C.4 (iii), is below 1 kHz, lying almost precisely within the pass-band of the level 6 decomposition, although it was not necessary to use this information when applying the technique, which can be regarded as genuinely generic in nature.

Comparing the performance of the signal processing strategies as recorded in Figures C.5 and C.6, it can be noticed that the sharpest distinction between the period before the injection and the start of the injection is when wavelet decomposition level 6 is used in conjunction with the normal spectral distance measure.

## C.6     CONCLUDING REMARKS

Data from the 1994 and 1995 bench mark tests were used to compare the performances of seven different signal processing strategies proposed for the detection of boiling or a sodium/water reaction in LMFBR. Five of these strategies are discussed in this case study and the other two are detailed in case study A. The general signal processing strategy used in all cases relies on the signals from the normal background noise and the fault being additive, which gives rise to changes in the signal model in the time, frequency or probabilty domain. Two of the specific signal processing strategies, mean square prediction error (MSPE) and distance-in-parameter-space, are derived from an autoregressive model of the process, whereas the rest are implemented in the frequency domain using either global spectral distance measures or more particular spectral measures used in conjunction with wavelet analysis.

The emphasis throughout the work reported in this case study (and Case Study A also) has been to make no assumptions about the nature of the fault to be detected other than the principle of the additive nature of the signals from a fault and the background noise. This was done to develop generic strategies that could be applied in a variety of industrial situations and to anticipate the detection of fault conditions not previously envisaged.

In specific industrial situations where precise details of signals associated with the onset of fault conditions are available, other non-generic stategies may yield better results. Case Study B provides one example of this and uses narrowband filtering with envelope detection. The example detailed in section 8.6 of Chapter 8 shows another application of narrowband filtering used with considerable success when precise knowledge of the spectral characteristics of the fault was available.

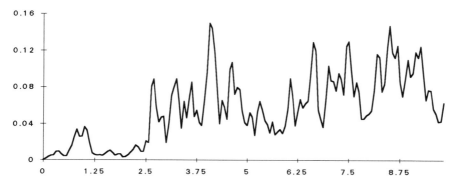

(i) Distance-in-parameter space vs. time for noise from injection 5 and evaporator background

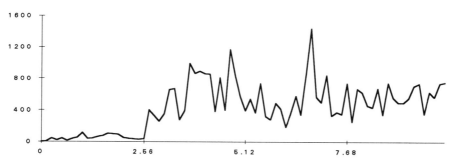

(ii) Direct spectral distance vs. time for noise from injection 5 and evaporator background

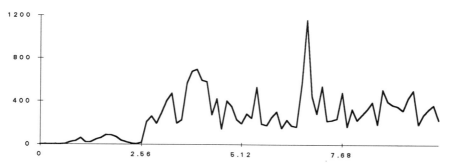

(iii) Normal spectral distance vs. time for noise from injection 5 and evaporator background

**Figure C.5** Comparison of three signal processing strategies using 1995 data

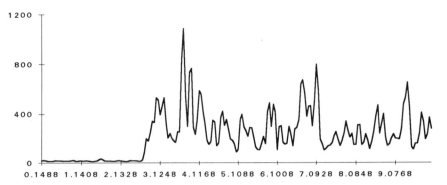

(i) Variance of the output of wavelet decomposition level 6 v time

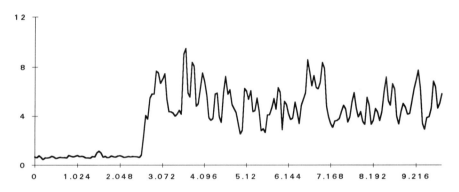

(ii) Direct spectral distance at the output of wavelet decomposition level 6

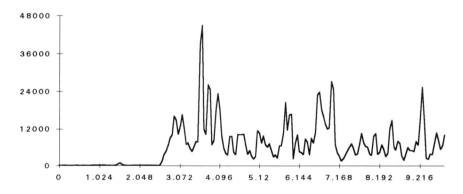

(iii) Normal spectral distance at the output of wavelet decomposition level 6

**Figure C.6** Comparison of variations with wavelets: injection 5 plus evaporator background using 1995 data

## C.7      CONCLUSIONS

- The energy associated with a test leak in the PFR evaporator lies predominantly in a frequency band below 1 kHz
- The mean square prediction error (MSPE) is the least effective method of fault detection and may only be used with confidence with signal-to-noise ratios greater than −12 dB
- The Distance-in-Parameter-Space Method is considerably superior to the MSPE method
- A combination of the use of wavelets with spectral distance measures is the most effective method for detecting the onset of an anomaly of the type that existed in the benchmark tests.

## C8      REFERENCES

Akaike, H., 1970, A new look at statistical model identification, *IEEE Transactions on Automatic Control,* AC-19, 716-723.

Black, J. L., Rofe, S. E. F., & Ledwidge, T. J., 1993 *IAEA Extended Co-ordinated Research Programme on Acoustic Signal Processing for the Detection of Boiling or Sodium/Water Reaction in LMFBR*: Report to a meeting of the research co-ordination meeting, Vienna, Austria, 9-10 December 1993.

Daubechies, I., 1988, Orthonormal Bases of Compactly Supported Wavelets, *Comm. in Pure and Applied Math.,* . 41, No. 7, 909-996.

Rofe, S. E. F., & Ledwidge, T. J., 1994, *IAEA Extended Co-ordinated Research Programme on Acoustic Signal Processing for the Detection of Boiling or Sodium/Water Reaction in LMFBR*: Report to a meeting of the research co-ordination meeting, Kalpakkam, India, 1-3 November 1994.

Rofe, S. E. F., & Ledwidge, T. J., 1995, *IAEA Extended Co-ordinated Research Programme on Acoustic Signal Processing for the Detection of Boiling or Sodium/Water Reaction in LMFBR*: Report to a meeting of the research co-ordination meeting, Obninsk, Russia, 11-14 July 1995.

Srinivasan, G. S., & Singh, O. P., 1990, 'New Statistical Features Sensitive to Sodium Boiling Noise', *Annals. of Nuclear. Energy* 17, No. 3, 135-138.

# Case Study D

## POWER STATION BOILER DYNAMICS

This case study is an edited version of a paper by T. M. Romberg

### OVERVIEW

This case study describes how measurements of the coal feed, feedwater flow, drum level, steam flow, TSV pressure error and generator power can be used to identify the dynamic behaviour of a 500 MW(e) drum boiler during full power operation under closed loop control. The fluctuations in these measurements (inherent noise) are analyses using multivariate time series techniques and a hydrodynamic model to identify the fundamental mode dynamics of the boiler.

It presents an overview of multivariate systems analysis theory, and shows how selected parameters of the hydrodynamic model are estimated using the measured correlation functions as reference descriptors. The sensitivity of the model results to changes in the subcooled boiling boundary, two-phase slip, two-phase friction multiplier and heat transfer correlation coefficients, is also discussed.

### D.1    INTRODUCTION

Boiling water reactors, fossil-fired boilers and other industrial two-phase heat transfer processes are normally operated under closed loop control to satisfy load demands imposed by the external system. In this mode of operation the process variables (coolant pressures, temperatures, flow rates, heat fluxes, etc.) fluctuate about the steady state level, particularly if the plant is used for peak load rather than base load operation. Attempts in the early 1960s to utilise this 'inherent noise' were encouraging, but more stringent control and safety regulations, coupled with significant developments in process instrumentation and computer technology in recent years, have led to a rapid growth in the number of investigations attempting to monitor and analyse noise signals for process performance, diagnostic analysis and surveillance, as well as the evaluation of mathematical models for process design and control studies. The inherent noise method is particularly relevant to the analysis of complex large scale commercial plant, where traditional methods for superimposing test signals are impractical during normal operation.

In contrast to most modelling studies, which use inherent noise techniques and autoregressive (AR) or autoregressive moving average (ARMA) parametric models to identify the process, the present study uses a distributed hydro-dynamic model derived from fundamental mass, energy and momentum conservation equations. Previous work (see Case Study E) has shown how the model and multivariate spectral analysis methods can be used to assess the two-phase flow stability characteristics of a boiling channel from measurements of the inherent hydrodynamic noise over a range of power levels. The aim of the present paper is to show how multivariate time series analysis and distributed modelling concepts can be applied to the hydrodynamic analysis of a commercial drum boiler. The paper presents an overview of multivariate process analysis theory, describes more recent modifications to the hydrodynamic model, and shows how they are used conjunctively to assess the hydrodynamic performance of a 500 MW(e) drum boiler from data logged during normal operation under feedback control. The correlation functions computed from the data are used as input perturbations for the model, and as reference descriptors for evaluating its accuracy in identifying the hydro-dynamic characteristics of the boiler. The sensitivity of the results to changes in the coefficients of the subcooled boiling, two-phase slip, friction multiplier and heat transfer correlations, is also discussed.

## D.2    MULTIVARIATE SYSTEM ANALYSIS THEORY

The hydrodynamic behaviour of the drum boiler is identified from the multiple input single output model depicted in Figure D.1. In the following sections, some basic definitions, conditions for identifying a process model and multi-variable equations relevant to the present investigation are derived. Readers are referred to Chapters 4 and 5 for further details.

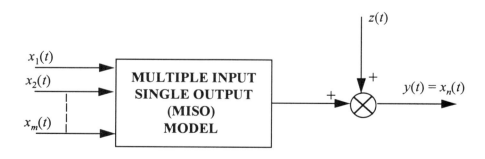

**Figure D.1** Multivariate model of a typical linear process

### D.2.1    Basic Definitions

A single-input-single-output (SISO) model in which the output is linearly related to a single input, is the basic building block for more general multivariate models. For a stationary bivariate (SISO) process, the cross-correlation function is defined in terms of the input variable $x(t)$ and the output variable $y(t)$ by the convolution equation:

$$R_{xy}(\tau) = E[\{x(t) - \bar{x}\}\{y(t + \tau) - \bar{y}\}], \qquad \ldots \text{(D.1)}$$

where $\tau$ is the time lag between the input and output variables, $E[\,]$ is the expected value operation and the mean values $\bar{x}$ and $\bar{y}$ are defined by $\bar{x} = E[x(t)]$ and $\bar{y} = E[y(t)]$ respectively. In the following analyses, the mean values are assumed, without loss of generality, to be zero. (This means that the mean values are removed in the pre-processing stage, that is, before computation of the correlation functions.) Similarly, the input and output auto-correlation functions are defined by:

$$R_{xx}(\tau) = E[x(t)x(t + \tau)] \qquad \ldots \text{(D.2)}$$

and

$$R_{yy}(\tau) = E[y(t)y(t + \tau)] \qquad \ldots \text{(D.3)}$$

### D.2.2    Minimum Least Squares Process Analysis

Consider the process with input and output realizations $x(t)$ and $y(t)$ respectively, and where the output has a superimposed error term $z(t)$ which contains a systematic component due to the inadequacy of the SISO linear process approximation, and a random component due to measurement errors and other controlled variables affecting the output. The output is calculated from the input $x(t)$ and the process weighting function $h(\tau)$ by the convolution integral:

$$y(t) = E[x(t - \tau)h(\tau)] + z(t) \qquad \ldots \text{(D.3)}$$

in which, for a physically realizable process,

$$h(\tau) = 0 \quad \text{for } \tau < 0 \qquad \ldots \text{(D.4)}$$

Substituting (D.3) in (D.1) and minimising the mean square residual error (see Chapter 4) yields the Wiener-Hopf integral equation:

$$R_{xy}(\tau) = E[R_{xx}(\tau - \lambda)\hat{h}(\lambda)] \qquad \ldots \text{(D.5)}$$

where $\hat{h}(\lambda)$ denotes the 'optimum' estimate of the process weighting function $h(\lambda)$ in the least squares sense. Note that equation (D.5) is equivalent to assuming that the input $x(t)$ and the residual error $z(t)$ are independent (uncorrelated); that is $E[x(t)z(t+\tau)] = 0$ for all $\tau$. However, if the input contains measurement or process noise such that $x'(t) = x(t) + u(t)$, then the input autocorrelation function $R_{xx}(\tau)$ becomes:

$$R'_{xx}(\tau) = R_{xx}(\tau) + R_{uu}(\tau)$$

and the optimum process weighting function $\hat{h}(\lambda)$ cannot be identified from the measurements except in special cases, e.g. where $x(t)$ and $z(t)$ have different bandwidths.

### D.2.3    Feedback Process Analysis

Most practical processes have inherent feedback or feedback controllers, whose action is to prevent non-stationary drift in the plant/process operating conditions. The analysis of closed loop processes poses special problems regarding their identifiability and the accuracy of the parameter estimates obtained (see Chapter 5), and cannot be identified by direct application of spectral methods (e.g. Chapter 4).

Consider the closed loop process depicted in Figure D.2, in which the measurements are $x(t)$ and $y(t)$, and $v(t)$ and $z(t)$ are extraneous noise sources superimposed on the input and output respectively. The convolution equations that relate the variables are:

Feedforward:  $y(t) = E[x(t-\tau)h(\tau)] + z(t)$                ... (D.6a)

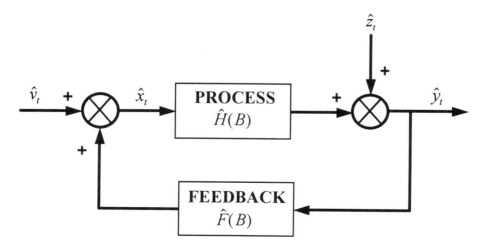

**Figure D.2** Block diagram of a closed loop process

Feedback: $\qquad x(t) = E[y(t-\tau)f(\tau)] + v(t) \qquad\qquad$ ... (D.6b)

The corresponding cross-correlation functions for the feedforward and feedback loops are respectively:

$$R_{xy}(\tau) = E[R_{xx}(\tau-\lambda)\hat{h}(\lambda)] \qquad \tau \geq 0$$
$$\qquad\quad = R_{xz}(\tau) \qquad\qquad\qquad\qquad \tau < 0 \qquad\qquad \text{... (D.7)}$$

and

$$R_{yx}(\tau) = E[R_{yy}(\tau-\lambda)\hat{f}(\lambda)] \qquad \tau \geq 0$$
$$\qquad\quad = R_{yv}(\tau) \qquad\qquad\qquad\qquad \tau < 0 \qquad\qquad \text{... (D.8)}$$

Equations (D.7) and (D.8) assume that $v(t)$ and $z(t)$ are 'white' noise sources such that $R_{xz}(\tau) = R_{yv}(\tau) = 0$ for $\tau \geq 0$. This assumption, plus the *proviso* that the two processes are suitably separated by process lags, enables the feedforward process weighting function $\hat{h}(\tau)$ to be identified from the cross-correlation estimates in the positive (+ve) lag domain, and the feedback controller/weighting function $\hat{f}(\tau)$ to be identified from the cross-correlation estimates in the negative (−ve) lag domain. In practice, this means the measurands must be non-deterministic and broadband in character, and narrowband filtering of the measurements in the pre-processing stage will increase the correlation between neighbouring estimates of the cross-correlation function and cause overlap at the zero lag ($\tau = 0$) axis.

### D.2.4    Multivariate Process Analysis

The bivariate concepts discussed in the previous sections are readily extended to the analysis of maultivariate linear processes. However, as discussed in Chapter 7, the general multiple-input-multiple-output (MIMO) process model cannot be identified in the minimum least squares sense unless certain constraints are placed on the residual noise matrix which effectively reduce the MIMO process to a set of multiple-input-single-output (MISO) sub-processes.

The output $x_n(t) = y(t)$ of a MISO process is a linear sum of the inputs $x_j(t)$ ($j = 1,m;\ n = m+1$) as defined by the convolution equation:

$$x_n(t) = E\left[ \sum_{j=1}^{m} x_j(t) h_{jn}(t+\tau) \right] + z(t) \qquad\qquad \text{... (D.9)}$$

where $h_{jn}(\tau)$ are the corresponding input-output process weighting functions. The cross-correlation between the *i*th input variable and the output is given by:

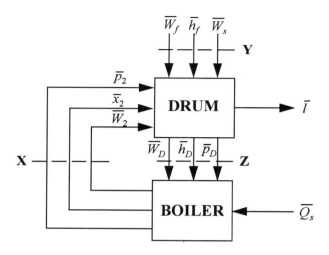

**Figure D.3** Block diagram of drum boiler model

$$R_{in}(\tau) = E[x_i(t)x_n(t+\tau)]$$

that is,

$$R_{in}(\tau) = \sum_{j=1}^{m} E[R_{ij}(\tau - \lambda)\hat{h}_{jn}(\lambda)] \qquad \dots \text{(D.10)}$$

which is the multivariate form of equation (D.5).

## D.3    HYDRODYNAMIC MODEL

The hydrodynamic model represents the drum and boiler dynamics by the transfer matrix 'blocks' depicted schematically in Figure D.3.

The boiler model is derived from the one-dimensional partial differential equations for the conservation of mass, energy and momentum in conjunction with empirical correlations to predict the onset of subcooled boiling and model the flow characteristics in the single phase (subcooled liquid) and two-phase (subcooled, saturated boiling) regions of the evaporator. The equations are perturbed in time, linearised and Laplace transformed to yield nonlinear ordinary differential equations which are integrated spatially using finite difference techniques.

The finite difference equations for the $j$th space node may be manipulated to give a transfer matrix equation of the form:

$$\begin{bmatrix} \overline{w}_j \\ \overline{h}_j \\ \overline{p}_j \end{bmatrix} = \begin{bmatrix} \overline{T}(w_j,w_{j-1}) & \overline{T}(w_j,h_{j-1}) & \overline{T}(w_j,p_{j-1}) \\ \overline{T}(h_j,w_{j-1}) & \overline{T}(h_j,h_{j-1}) & \overline{T}(h_j,p_{j-1}) \\ \overline{T}(p_j,w_{j-1}) & \overline{T}(p_j,h_{j-1}) & \overline{T}(p_j,p_{j-1}) \end{bmatrix} \begin{bmatrix} \overline{w}_{j-1} \\ \overline{h}_{j-1} \\ \overline{p}_{j-1} \end{bmatrix} + \begin{bmatrix} \overline{Q}_{1j} \\ \overline{Q}_{2j} \\ \overline{Q}_{3j} \end{bmatrix} \overline{Q}_s \dots \text{(D.11)}$$

which relates the frequency dependent mass flow ($\overline{w}$), specific enthalpy ($\overline{h}$) and pressure ($\overline{p}$) at the inlet ($j-1$) and exit ($j$) of the node by a ($3\times3$) transfer matrix $\overline{\mathbf{T}}_j$ and a ($3\times1$) column vector $\overline{\mathbf{Q}}_j$ to account for heat source ($\overline{Q}_s$) perturbations. Equation (D.11) may be rewritten as:

$$\mathbf{X}_j = \mathbf{T}_j\,\mathbf{X}_{j-1} + \overline{\mathbf{Q}}_j\,\overline{Q}_s \qquad\qquad \text{... (D.12)}$$

If the boiler circuit is subdivided into $N$ space nodes, then the transfer matrix blocks are cascaded $N$ times to yield the overall input-output transfer matrix equation:

$$\mathbf{X} = \mathbf{HZ} + \mathbf{P}\overline{Q}_s \qquad\qquad \text{... (D.13)}$$

where

$$\mathbf{H} = \mathbf{H_N} = \prod_{j=1}^{N}\mathbf{T}_j = \mathbf{T_N H_{N-1}} \qquad\qquad \text{... (D.14)}$$

$$\mathbf{P} = \mathbf{P_N} = \mathbf{Q_N} + \mathbf{T_N P_{N-1}} \qquad\qquad \text{... (D.15)}$$

and $\mathbf{Z}$ is a ($3\times1$) column vector of the mass flow, enthalpy and pressure perturbations at the downcomer inlet. The continuities in flow conditions at nodal boundaries with local area changes/restrictions and at the boiling boundary are conserved by coupling equations.

The drum model is derived from the macroscopic equations for the conservation of mass, energy and volume in the separate steam and water phases and the total system. The equations are perturbed, linearised and Laplace transformed to yield a matrix equation of the form:

$$\mathbf{Z} = \mathbf{BX} + \mathbf{CY} \qquad\qquad \text{... (D.16)}$$

where     $\mathbf{Z'}$ = vector of downcomer inlet perturbations = $[\overline{w}_D,\ \overline{h}_D,\ \overline{p}_D]$
            $\mathbf{X'}$ = vector of evaporator exit perturbations   = $[\overline{w}_2,\ \overline{h}_2,\ \overline{p}_2]$
            $\mathbf{Y'}$ = vector of feedwater/steam perturbations = $[\overline{w}_F,\ \overline{h}_F,\ \overline{w}_S]$

and $\mathbf{B}$ and $\mathbf{C}$ are ($3\times3$) matrices similar in form to the $\mathbf{T}_j$ matrix in equation (D.11).

Insertion of equation (D.16) in (D.13) yields

$$\mathbf{X} = \mathbf{G}^{-1}[\mathbf{HCY} + \mathbf{P}\overline{Q}_s], \qquad\qquad \text{... (D.17)}$$

where the ($3\times3$) characteristic matrix

$$\mathbf{G} = [\mathbf{I} - \mathbf{HB}] \qquad\qquad \text{... (D.18)}$$

Note that the drum boiler is given by the locus of the determinant:

$$\Delta = |\mathbf{G}| > 0 \qquad \qquad \text{... (D.19)}$$

in the complex plane (Argand diagram); that is, the determinant must not encircle the origin.

The drum level perturbation is calculated from the vector equation:

$$\bar{l} = \mathbf{EX} + \mathbf{FY} \qquad \qquad \text{... (D.20)}$$

where $\mathbf{E}$ and $\mathbf{F}$ are $(1 \times 3)$ row vectors. Inserting equation (D.17) into (D.20) yields:

$$\bar{l} = \left[ \mathbf{EG^{-1}HC} + \mathbf{F} \right] \mathbf{Y} + \mathbf{EG^{-1}P}\bar{Q}_s = \mathbf{U'W} \qquad \qquad \text{... (D.21)}$$

where $\mathbf{U'}$ = row vector of inputs = $[\bar{w}_F, \ \bar{h}_F, \ \bar{w}_S, \ \bar{Q}_s]$, and $\mathbf{W}$ is a $(4 \times 1)$ column vector of input-output transfer functions. Inverse Laplace transformation of equation (D.21) yields (cf. equation (D.9)):

$$l(t) = u_s(t) = E\left[ \sum_{j=1}^{4} u_j(t - \lambda) w_{j5}(\lambda) \right] \qquad \qquad \text{... (D.22)}$$

The cross-correlation between the $i$th input $u_i(t)$ and the output (drum level, $u_s(t)$) is given by:

$$R_{i5}(\tau) = E[u_i(t) u_s(t + \tau)] \qquad i = 1,4$$
$$= \sum_{j=1}^{4} E[R_{ij}(\tau - \lambda) \tilde{w}_{j5}(\lambda)] \qquad \qquad \text{... (D.23)}$$

## D.4 BOILER PLANT AND DATA ACQUISITION

### D.4.1 Description of Plant

The drum boiler investigated in this study is a 500 MW(e) unit situated at Wallerawang in New South Wales, Australia. A simplified schematic of the boiler and its associated turbine-generator units is shown in Figure D.4. The primary circuit consists of five interacting sub-processes: a steam drum, downcomers, circulating pumps, evaporator (water walls) and risers. External connections to and from the turbines via the spray superheaters and economiser have not been included in the present model.

Subcooled water flows from the drum through vertical downcomers to four circulating pumps, which pump it to lower headers at the base of the furnace.

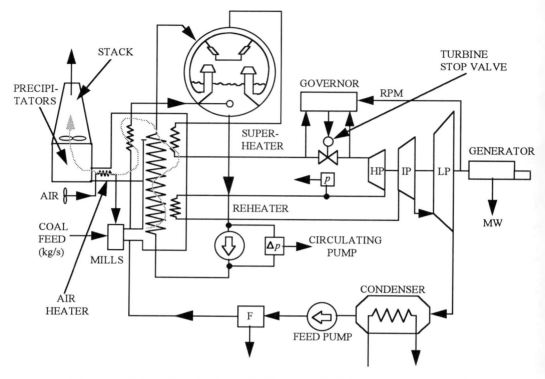

**Figure D.4** Simplified schematic diagram of Wallerawang power station

The lower headers distribute the flow through about 1200 vertical pipes spaced around the outer and central walls of the evaporator. These pipes (or channels) are connected to upper headers which, in turn, are connected to the steam drum by riser pipes. The two-phase (saturated water/steam) mixture enters the upper part of the drum, and is directed to turbo separators by a cylindrical shell mounted inside the drum. The steam leaving the turbo separators passes through a series of plate dryers before entering the spray superheaters. The saturated water mixes with the subcooled feedwater entering the economiser through horizontal holes in the sparge tubes above each of the downcomers.

### D.4.2    Data Acquisition

Measurements of the variations in the coal feed to the pulverising mills, feedwater flow, steam flow to the turbine, pressure error at the turbine stop valve (TSV) and drum level were recorded on the boiler during full power operation in a boiler-follow control mode; that is, the coal feed-to-mills is controlled to minimise the error between the upstream TSV pressure and its set point of 15.8 MPa. The analogue signals obtained from the station control panels were recorded on magnetic tape using a 7-track Ampex FR1300 taper

**Table D.1** Steady state data from computer log

| | |
|---|---|
| Drum pressure | 16.45 Mpa |
| Drum level | 48.1 cm (53.3 cm datum) |
| Economiser inlet temperature | 187 °C |
| Economiser outlet temperature | 268 °C |
| Feedwater flow | 389 kg/s |
| Downcomer inlet temperature | 349 °C |
| Circulating pump $\Delta p$ | 228 kPa |
| Coal feed to mills | 47.3 kg/s |
| Generator power | 480 MW(e) |

recorder. The tape records were subsequently digitised at 1.6 second intervals to give 3600 data points/record, and these were stored on the disc units of an IBM3033S mainframe computer. The station computer was also programmed to print a snapshot log at one minute intervals of the drum pressure and level, feedwater flow, economiser inlet and outlet temperatures, coal flow, pressure drop across the circulating pumps, generator power and downcomer inlet temperatures. A five channel pen recorder was also used to monitor in parallel selected signals recorded on tape.

### D.4.3    Steady State

The data from the snapshot log were averaged over the recording period (2 hours) to obtain the steady state information given in Table D.1.

Energy balances based on this data yielded economiser, evaporator and superheater powers of 145, 543 and 325 MW respectively. The recirculation flow calculated from the pump characteristic was 2243 kg/s (4 pumps).

This basic data and the primary circuit dimensions supply the input data for the hydrodynamic model computer code. The primary circuit steady state conditions were matched by adjustment of the loss factor for the gags (orifices) inserted at the inlet of the water wall tubes. The computations yielded an evaporator exit quality of 26.5%, which corresponds to a steam flow of 594 kg/s. However, the measured steam flow to the superheater was 389 kg/s, which indicates that 205 kg/s of saturated steam was condensed in raising the feedwater temperature to almost saturation, as confirmed by the downcomer inlet temperatures.

**Figure D.5** Multivariate model of the drum boiler

## D.5      BOILER IDENTIFICATION

### D.5.1    Data Analysis and Results

The correlation functions for the MISO models discussed in sections D.2 and D.3 were computed from the digitised data. The power spectral density estimates for the five records (feedwater flow, steam flow, coal feed, TSV pressure error, drum level) were calculated after broadband digital filtering to remove the d.c. (mean) values and frequencies above the Nyquist frequency (0.3125 Hz). The spectral density estimates and dominant amplitudes in the 0.001–0.1 Hz frequency range, and the digital filter settings, were adjusted to bandpass information in this range.

The correlation functions of the MISO model relating feedwater flow, steam flow, coal feed (inputs) to drum level (output), Figure D.5, were computed as outlined in section D.2. The maxima/minima of the various combinations of cross-correlation functions and their associated lag in seconds are summarised in Table D.2, and yield the following observations:

- the coal feed and TSV pressure error are highly correlated (−0.92) due to feedforward control action;

**Table D.2** Normalised cross-correlation maxima/minima

| Cross-correlation | | Max/Min | Lag (s) |
|---|---|---|---|
| Coal feed | → Feedwater flow | +0.21 | + 4.8 |
| | → Drum level | +0.49 | +20.8 |
| | → TSV pressure error | −0.92 | + 1.6 |
| | → Steam flow | −0.84 | + 8.6 |
| Feedwater flow | → Drum level | −0.51 | −19.2 |
| | → TSV pressure error | +0.35 | −36.8 |
| | → Steam flow | +0.28 | −32.0 |
| Drum level | → TSV pressure error | −0.50 | −19.2 |

- the TSV pressure error and the steam flow, as inferred from the pressure after the HP turbine stage (Figure D.4), are also highly correlated (+0.93);
- the steam flow and coal feed are also highly correlated (−0.84) by the process and control mechanisms; the lag of only 8 seconds suggests the dominance of the latter mechanism;
- hence the steam flow is not an independent input, and is eliminated from the MISO model in the boiler identification;
- the feedwater flow is predominantly correlated with the drum level via the control loop, as indicated by the lead (negative lag) of 19.2 seconds.

## D.5.2    Identification Procedure

The aim of the identification procedure is to determine the model weighting functions, $\hat{w}(\tau)$ in equation (D.23), which are compatible with the drum boiler weighting functions $\hat{h}(\tau)$, equation (D.10), by minimisation of the root mean square (RMS) errors between the equivalent cross-correlation functions.

The hydrodynamic model result at the same steady state operating level (as logged) is computed from:

$$R_{in}(\tau) = \sum_{j=1}^{m} E[\hat{R}_{ij}(\tau - \lambda)\tilde{w}_{jn}(\lambda)] \qquad \ldots \text{(D.24)}$$

where $i = 1, m,$ and $\hat{R}_{ij}(\tau)$ denotes the cross-correlation functions computed from the measurements, and which are used to perturb the drum boiler model. The RMS error between the predicted input-output cross-correlation $R_{in}(\tau)$ is given by:

$$\varepsilon_{RMS} = E[\hat{R}_{in}(\tau) - R_{in}(\tau)]^2 \qquad \ldots \text{(D.25)}$$

Substitution of equations (D.10) and (D.23) into (D.25) yields

$$\varepsilon_{RMS} = \left\{ E[E[\hat{R}_{in}(\tau - \lambda)\{\hat{h}_{jn}(\lambda) - \tilde{w}_{jn}(\lambda)]]^2 \right\}^{\frac{1}{2}} \qquad \ldots \text{(D.26)}$$

Thus, the RMS error tends to zero as $\hat{h}_{jn}(\lambda) \rightarrow \tilde{w}_{jn}(\lambda)$.

## D.5.3    Coal Feed→Drum Level Analysis

The drum boiler model of the coal feed drum→level cross-correlation at the same steady state operating level is computed from:

$$R_{13}(\tau) = E[\hat{R}_{11}(\tau - \lambda)\tilde{w}_{13}(\lambda) + \hat{R}_{12}(\tau - \lambda)\tilde{w}_{23}(\lambda)] \qquad \ldots \text{(D.27)}$$

where the subscripts 1, 2, 3 denote the coal feed, feedwater flow (inputs) and drum level (output) respectively, and $-\tau_m \leq \tau \leq \tau_m$ ($\tau_m$ = maximum lag).

In utilising equation (D.27), the model requires power perturbations in the evaporator water walls as an input, and so some assumptions are necessary regarding the mill to furnace transfer function. The internal grinding and combustion processes are extremely complex to model, and in this analysis it is assumed that the coal feed and water wall power can be represented by a simple transfer function model of the form:

$$H_{MF}(s) = K_{MF}\, e^{-s\tau_{MF}} \qquad \qquad \text{... (D.28)}$$

where $\tau_{MF}$ is the mill/furnace (MF) transport delay, $s$ is the Laplace operator, and $K_{MF}$ is the static gain. The static gain $K_{MF}$ denotes a coal feed (CF) →evaporator power (EP) ratio, which can be approximated by:

$$\frac{\sigma_{EP}}{M_{EP}} = K_{MF}\frac{\sigma_{CF}}{M_{CF}} \qquad \qquad \text{... (D.29)}$$

where $M_{EP}$, $M_{CF}$ and $\sigma_{EP}$, $\sigma_{CF}$ are, respectively, the evaporator power and coal feed mean values and standard deviations.

In order to avoid a non-minimum phase identification problem, the mill/furnace transport delay $\tau_{MF}$ must be estimated first. The change in coal feed→drum level RMS error, equation (D.25), as a function of $\tau_{MF}$ is given in Figure D.6a, and has a minimum at 19.2 seconds. The reference model for these calculations assumed that:

a)  subcooled boiling,
b)  homogeneous flow,
c)  no slip between the steam/water phases, and
d)  no thermal storage in the water wall tubes.

Similarly, estimation of $K_{MF}$ by minimisation of the RMS error yields an 'optimum' value of 0.3 (Figure D.6b).

Relaxation of the above thermal hydraulic assumptions by varying the parameters of the subcooled boiling, two-phase friction, slip and heat transfer correlations over physically realistic ranges, resulted in only marginal changes (<4%) in the minimum RMS error of 0.214, calculated in the range $-\tau_m \leq \tau \leq \tau_m$. The two-phase distribution parameter $(C_0)$ and the weighted mean drift velocity $(V)$ in the slip correlation had most influence on the RMS error. Increasing the distribution parameter from its homogeneous value $(C_0 = 1)$ increased the RMS error. However, decreasing the weighted mean

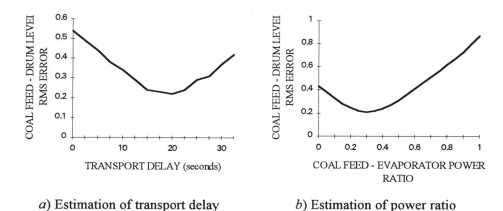

*a*) Estimation of transport delay      *b*) Estimation of power ratio

**Figure D.6** Estimation of mill/furnace transport delay and power ratio

drift velocity resulted in a minimum RMS error at $V = -1$ m/s. These values suggest that the two-phase flow in the water walls was probably annular, and the steam phase underwent some deceleration in the upper headers and horizontal riser tubes. Some improvement in the sensitivity of the RMS error could be achieved by invoking the physical realizability condition, equation (D.4), and confining the range to $0 \leq \tau \leq \tau_m$.

The comparison between the calculated and measured cross-correlation functions obtained by this identification procedure is shown in Figure D.7. The process (positive lag) results are in good agreement overall, although further refinements may produce marginally better agreement in the region of the peak

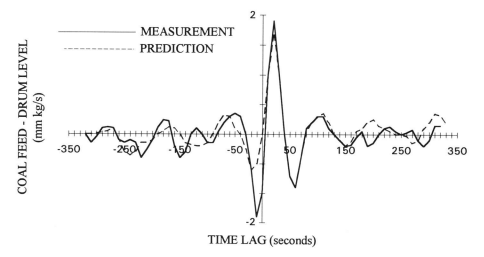

**Figure D.7** Comparison of coal feed→drum level cross-correlations

response. The negative lag (feedback) results agree less favourably, which is consistent with the open loop structure of the drum boiler model that is, external feedback loops via the turbine/condenser circuits have not been included.

### D.5.4     Feedwater Flow→Drum Level Analysis

The drum boiler model for the feedwater flow→drum level cross-correlation function is computed from the equation:

$$R_{23}(\tau) = E[\hat{R}_{21}(\tau - \lambda)\tilde{w}_{13}(\lambda) + \hat{R}_{22}(\tau - \lambda)\tilde{w}_{23}(\lambda)] \qquad \text{... (D.30)}$$

where the subscripts denote the same variables as in equation (D.27). The result obtained using the same steady state and parameter estimates as computed for the coal feed→drum level cross-correlation function, Figure D.7, is given in Figure D.8. The relatively good agreement, particularly in the positive lag domain, shows that the fundamental mode dynamic characteristics of the drum boiler are adequately described by the model. As previously, the negative lag (feedback) results are influenced by the action of the drum level control loop on the boiler operation.

### D.6     CONCLUSIONS

This case study has demonstrated how multivariate signal analysis techniques may be used to assess the thermal-hydraulic performance of a power station

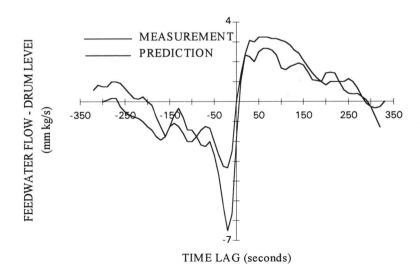

**Figure D.8** Comparison of feedwater flow→drum level cross-correlations

boiler from data logged under closed loop control during normal operations. Cross-correlation functions computed from measurements of the variations in coal feed, feedwater flow, drum level, TSV pressure error, steam flow and generator power recorded on a 500 MW(e) coal-fired drum boiler operating in boiler follow control mode, showed that:

1) the drum level was predominantly caused by coal feed and feedwater flow variations;
2) the computed correlation functions can be used to evaluate the heat transfer and fluid flow parameters of a distributed drum boiler model;
3) the coal feed-drum level and feedwater flow drum level cross-correlation function estimates computed from the measurements were in good agreement with those calculated by the hydrodynamic model after unknown mill/furnace parameters were estimated using minimum least squares techniques;
4) changes in the heat transfer and fluid flow parameters of the model over physically realistic ranges had only marginal affect (<4%) on the RMS error between the measured and calculated cross-correlation functions;
5) the weighted mean drift velocity between the steam and water phases of the two-phase slip correlation had most influence on the RMS error;
6) the optimised values suggest that the two-phase flow structure in the water walls was probably annular, and the steam phase decelerated in the upper headers and horizontal riser tubes;
7) the measurements were influenced by the closed loop/recycle circuits and boiler control action, as shown by the values of the cross-correlation ordinates at zero lag and negative lag domains;
8) from an identification viewpoint, the feedforward and feedback processes were, in this application, able to be identified from the computed cross-correlation estimates.

## REFERENCE

Romberg, T. M., 1984, Identification of Two-phase Hydrodynamic Processes from Normal Operating Data with Application to a Drum Boiler, *Multi-Phase Flow and Heat Transfer III. Part B: Applications* edited by T. N. Veziroglu and A. E. Bergles, Elsevier Science Publishers B.V., Amsterdam, 883-898.

# Case Study E

## SMELTING FURNACE DYNAMICS

This case study is an edited version of a paper by T. M. Romberg, M. R. Davis and R. W. Harris

### E.1    CONTEXT

A collaborative research program was set up between the CSIRO Division of Mineral Engineering and the Sulphide Corporation Pty Ltd to study the operation of the Imperial Smelting Furnace (ISF) connected to selected tuyere pipe tappings, and vibration transducers attached to the tuyere flanges adjacent to the furnace wall. A subsidiary investigation was also undertaken to monitor the vibration levels at three locations on the entrance of the condenser offtake roof, to detect the build-up of the zinc curtain that occurs during normal operation.

Vibration signals have been used extensively in a wide variety of industries to diagnose and identify acoustic sources and to analyse the response of structural components in fluid systems, mechanical systems and buildings. Much of the emphasis of the research in fluid systems has been oriented towards understanding the mechanisms of flow-induced vibrations and the structural integrity of the system components and containment, with the principal aim of developing an improved system design in which fatigue failure of the structural components is minimised.

By contrast, the present investigation aims to evaluate the diagnostic benefits of using vibration signals to monitor the thermal-hydraulics of smelting processes. The dominant source of acoustic energy emanating from the smelting process is caused by the turbulent zone adjacent to the tuyere nozzles (or raceway activity), which is dependent on operational factors such as:

*a)*  individual tuyere flow;
*b)*  bustle main pressure/total blast volume;
*c)*  the level of slag in the hearth, and hence the degree of interaction between the raceway gases and the slag;
*d)*  the combustion of coke in the raceway, the flame phenomena resulting from this, and the flow of gases into the packed bed;
*e)*  the physical movement of coke particles in the raceway;
*f)*  the size of the raceway.

Operational experience indicates that the most important factors are the individual tuyere flow and the degree of slag pool interaction. The initial phase of this project aims to investigate how the conditioned vibration signals correlate with the tuyere pressure measurements, and how they can be used to evaluate the smelting performances of the furnace.

## E.2    OBJECTIVES

The specific objectives of the initial investigation were to determine:

1) the degree of correlation between conditioned vibration signals and tuyere pressure measurements;
2) the degree of correlation between the conditioned structural vibration signals and selected process variables (e.g. bustle main pressure) logged during normal operation;
3) the degree of azimuthal interactions around the furnace by correlation between adjacent vibration signals and pressure measurements;
4) the sensitivity of vibration signals to changes in offgas turbulent flow due to build up of a zinc 'curtain' on the furnace offtake roof at the condenser entrance.

The first objective is intended to establish whether conditioned vibration signals are suitable for monitoring the thermal processes within the furnace. Operational experience indicated that the tuyere pressures are sensitive to the changing smelt characteristics, and are a useful datum for comparison purposes.

Having established that appropriately conditioned vibration signals can monitor changing smelt characteristics, the second and third objectives are intended to determine whether they can also track gross movements in ISF conditions and also detect azimuthal interactions around the furnace. This latter information is relevant for two and, possibly, three dimensional modelling studies.

The fourth objective is intended to develop an on-line technique for monitoring the curtain build-up on the roof of the furnace offtake, so that furnace operation and plant design can be improved.

## E.3    WORK PROGRAM

The work program involved the following tasks:

1) the installation of piezoelectric vibration and pressure sensors at four locations around the furnace;
2) the installation of vibration sensors on the furnace offtake roof stiffeners near the condenser entrance;

3) the acquisition of data from these sensors using a fourteen track magnetic tape recorder;

4) the acquisition of 15 selected plant variables at five minute intervals using a data logger and floppy disc storage under computer control;

5) on-line analysis of the vibration and tuyere pressure data using a Hewlett-Packard spectrum analyser, and the storage of selected spectra on floppy disc under computer control;

6) off-line multivariate time series analysis of the recorded data;

7) the reporting of significant results and conclusions, including recommendations for more comprehensive studies.

## E.4      INSTRUMENTATION AND INSTALLATION

The pressure transducers and piezoelectric accelerometers were mounted respectively on the pressure tappings and tuyere flanges at key locations around the furnace, and on the RSJ stiffeners of the condenser offtake roof, Figure E.1. Each sensor and its electrical connections to recording equipment in the control room are discussed in turn.

### E.4.1     Pressure Transducers

The pressure transducers were Kistler model 7001 units which use a piezoelectric element to convert pressure to electrical charge. The transducers were encased in special housings and were attached to existing pressure tappings on the horizontal sections of the tuyere pipes.     The pressure transducers had a nominal sensitivity of 74 pC/at, and were connected by

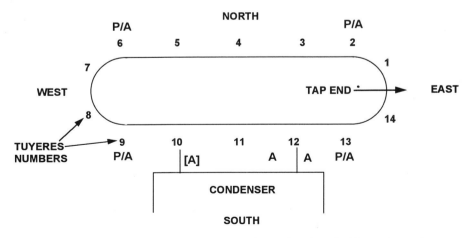

LEGEND:  **P** = piezoelectric pressure transducer   **A** = piezoelectric accelerometer

**Figure E.1** ISF pressure and accelerometer measurement points microdot coaxial cables to charge s sensitive amplifiers (Bruel & Kjaer model 2624, gain = 10 mV/pC) mounted nearby, and then via BNC coaxial cables to instrumentation amplifiers (PAR 113; gain=10, filter bandpass=0.03 Hz to 10 kHz) located in the control room.

### E.4.2    Vibration Transducers (Tuyeres)

The vibration transducers were Bruel & Kjaer accelerometers (model 4368) which use a piezoelectric element to convert the acceleration of an inertial mass into electrical charge. The transducers were attached to the flange at the bottom of the tuyere using magnetic bases. At one stage in the investigation two accelerometers were placed on the RSJ opposite tuyere 13. The sensitivity of the units was nominally 10 pC/g, and they were connected by microdot coaxial cables to charge sensitive amplifiers (Bruel & Kjaer model 2624, gain=10 mV/pC) located nearby, and then via BNC coaxial cables to instrumentation amplifiers (PAR 113; gain=10, filter bandpass settings=0.03 Hz to 10 kHz) located in the control room.

### E.4.3    Vibration Transducers (Condenser Roof)

The vibration transducers were Bruel & Kjaer accelerometers model 43687 and they were attached to an RSJ which was at the roof position near where the zinc curtain was known to form. The accelerometers were attached as shown in Figure E.1. The accelerometers were connected by microdot coaxial cable to charge sensitive amplifiers (Bruel & Kjaer model 2628; gain settings=10 mV/g, bandpass filter settings=0.3 Hz to 30 kHz) located nearby on the tuyere floor. The outputs of these amplifiers were connected by BNC coaxial cables to the tape recorder located in the control room.

### E.5    DATA ACQUISITION AND ANALYSIS

### E.5.1    Data Logging of Plant Variables

Selected process variables normally monitored for control of the ISF were digitised from their analogue signals at five minute intervals with a Hewlett Packard data logger (model 3421A) interfaced via a HP-IL bus to a Hewlett Packard HP-85 computer system, which was also interfaced via a HP-IB bus to a 9121 flexible disc drive for storage on magnetic disk, see Figure E.2. A list of the variables monitored and their respective ranges, is given in Table A.1, Appendix A.

### E.5.2    Data Recording

The analogue signals from the pressure transducers and the accelerometers were recorded on a 14 track instrumentation tape recorder TEAC model SR-50. The recorder channel numbers for each transducer are given in Table B.1, Appendix B.

A difficulty arose in determining the number of tapes required for the test, since this was fixed by the upper frequency of the pressure and vibration signals, which was unknown.   It was assumed prior to the test that the significant information would have an upper frequency of 1 kHz, and although this proved to be correct for the pressure signals, it was not correct for the vibration signals, which had an upper frequency of 10 kHz.   It became impracticable, therefore, to record for the entire time of the test (five days) as each tape lasted only 30 minutes, and 240 reels of magnetic tape would have been required instead of the 30 tapes purchased for the project.   Subsequent analysis of the tapes would also have been prohibitive, and beyond the scope of this feasibility study.

Selected periods of operation were consequently chosen to ensure that signals arising from important furnace conditions and operational phases were samples.

### E.5.3    On-Line Time and Frequency Analysis

On-line time and frequency domain analyses were performed to obtain an immediate indication of the sensitivity of the signals to variations in furnace performance and operation. A schematic diagram of the equipment connections is shown in Figure E.2. Priority was given to the acquisition and storage of the process data being monitored by the HP-3421A data logger every five minutes; this was stored automatically on disc#0 of the HP-9121 dual floppy disc unit. Ancillary time/spectral information could also be stored on disc#1 during this time interval by pressing programmed function keys on the HP-85 computer.

#### E.5.3.1    Filtered Signal Analysis

A single channel spectrum analyser (Spectral Dynamics model SD301B), which uses a mixture of analogue and digital techniques to compute the spectra, was employed as a narrowband filter for the accelerometer signal, so that variations of the signal amplitude in a sensitive frequency band could be observed. The filtered output from the Spectral Dynamics unit was passed to a true RMS voltmeter (Hewlett Packard model 3400A) and the output from this meter was displayed on a two pen chart recorder (Brush model 220) and also recorded on channel 13 of the tape recorder. The pressure transducer signal in the same tuyere was displayed on the second channel of the Brush recorder after it had

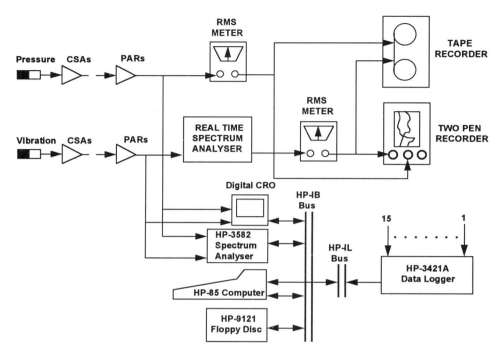

**Figure E.2** Schematic for data logging and on-line signal analysis

been conditioned by a second RMS voltmeter. The conditioned pressure signal was also recorded on channel 12 of the tape recorder.

### E.5.3.2 CRT Display

The same pressure and vibration signals were also connected to a dual channel digital oscilloscope (Hameg model HM208) for viewing and digitising selected signals. These could be transferred via the HP-IB bus to the HP-85 computer for storage on disc if required.

### E.5.3.3 Dual Channel Spectral Analysis

The same transducer signals were also connected to a dual channel FFT spectrum analyser (Hewlett Packard model 3582A), which computes the power spectral density estimator (PSDs) for individual signals and the cross-spectral density estimates (CSDs) between pairs of signals so that the frequency characteristics of the signals and their inter-relationship can be determined (see Chapter 4). The spectral estimates were displayed on a screen, and selected results were transferred to the HP-85 computer system for storage on a floppy disc. These included the two power spectral densities, the coherence and the phase estimates. The on-line spectral analysis allowed variations in the spectral characteristics of the signals to be immediately observed and related to any changes in the operation of the furnace, as discussed in the following section.

### E.5.4    Data Re-recording and Analysis

Selected segments of the signals recorded on the TEAC 14-Track instrumentation tape recorder were re-recorded on to an AMPEX FR700 7-track tape recorder in the Noise Analysis Laboratory at Lucas Heights. This was necessary for long-term storage and analysis because the TEAC recorder was rented from Tech Rentals for a limited period of time. Only seven of the twelve channels could be re-recorded simultaneously from the TEAC to the AMPEX, and it was decided that these should be the vibration and pressure signals in the tuyeres 2, 9 and 13. Re-recording the 30 TEAC tapes would have been prohibitive in time, and so the metallurgists at Sulphide Corporation were asked to nominate the TEAC tapes which were most important for detailed signal analysis. Re-recording was done to analyse significant events such as tuyere punching, blast volume transients, blown seals, etc. as well as normal plant operation.

### E.6       RESULTS AND DISCUSSION

### E.6.1    Gross Plant Movements

A typical record of the digitised ISF data logged during the test is given in Figure E.3, which shows the condenser off-take pressure (kPa) during the test period. The large dip in pressure commencing at 25 hours after the start of logging was due to a blown underflow seal in the tap discharge, and the reduction of the blast volume flow from 35,000 m$^3$/h to 5000 m$^3$/h. These digitised records are used for comparison with the peak amplitude results obtained as discussed in the next section.

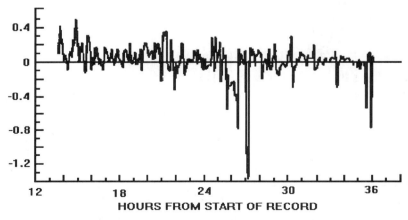

**Figure E.3** Condenser offtake pressure (kPa)

*E.6.1.2    Peak Spectral Amplitudes*

Towards the latter stages of the test period, it was noted that certain peak amplitudes of the vibration signal PSDs in the 6-9 kHz frequency range, appeared to follow gross movements in ISF operation. The peak amplitudes on tuyeres 9 and 13 were closely monitored in the 22 hours before the ISF was shut down. A typical evolutionary analysis of tuyere 9 vibration spectra for the loss of slag flow period is shown in Figure E.4. It is evident that gross reductions in the bustle main pressure are mirrored by the corresponding movements in the peak amplitudes. It is postulated that the primary mechanism relating tuyere vibration to the bustle main pressure is the turbulent energy of the raceway activity. The peak spectral amplitudes exhibited some 'fine structure' transient effects not directly related to bustle main pressure. This suggests the variation may be due to an internal furnace source, which could be the result of:

*a)*   changes in the liquid levels in the hearth of the furnace caused by a non-uniform pressure distribution throughout the furnace;

*b)*   slumping of the furnace burden in this region as the bustle main pressure decreased;

*c)*   a marked change in the blast distribution between individual tuyeres resulting from a non-uniform burden distribution;

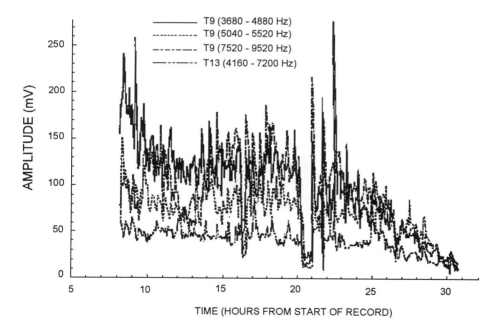

**Figure E.4** Comparison of vibration peak spectral amplitudes for tuyeres 9 and 13

*d*)  surges of lead and slag flowing through the furnace, resulting in a 'wave' of material passing in front of the tuyere;

*e*)  a temporary imbalance between the furnace pressure, liquid level in the hearth and the atmosphere.

In the last case, the flow of lead and slag is maintained by the pressure differential between the furnace and the atmosphere. Under normal conditions, a continuous liquid flow results. Any change in the pressure balance thus affects the liquid flow from the furnace. However, tuyere 9 also showed strong variations in peak spectral amplitude level which were not reflected in bustle main pressure variations. It is possible that raceway activity was affected by longitudinal slag height variations or changes in liquid drainage characteristics within the furnace.

### E.6.2    Tuyere Raceway: Evolutionary Spectral Analysis

Evolutionary spectral plots of tuyere raceway activity were computed for:

- *tuyere punching operations*, in which a long metal rod is inserted down the tuyere to clear away accumulated dross and smelt from the tuyere nozzle, which may block the air flow into the furnace;
- *a blown seal incident*, in which hot air is inadvertently discharged from the furnace; and
- *a loss of slag flow incident*, in which the slag flow orifice becomes blocked during tapping.

A typical evolutionary spectral plot is shown in Figure E.5, which shows the evolutionary spectral record for a loss of slag flow incident as recorded by the

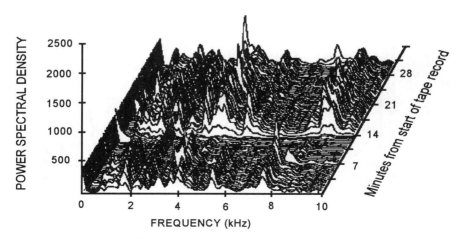

**Figure E.5**  Loss of slag flow: evolutionary spectra of tuyere 9 vibration

vibration sensor mounted on tuyere 9. Note the significant change in spectral characteristics at the 14 minute time mark, particularly the spectral amplitudes in the 4, 6 and 8 kHz frequency bands. The record from time zero to 6.00 minutes corresponds to a period of approximately two hours before a loss of slag flow, while the tape record from 6.00 to 13.00 minutes covers the period following the loss of slag flow, when the blast volume flow had been decreased, and that there is little activity in these frequency bands. The tape record between 13.00 and 31.00 minutes shows the period during and after the blast volume flow was restored to 36,000 m³/h. Note that there is now significant activity in the frequency bandwidths of interest.

### E.6.3    Tuyere Pressure and Vibration Analysis

The relationship between tuyere 13 pressure and vibration is given in Figure E.6, which gives a comparison of their PSDs. Both traces show significant peaks in the 700-950 Hz frequency bandwidth, but this was not reflected in the computed coherence estimates. One of the following conclusions can be drawn from this result:

1) The tuyere pressure reflects raceway activity but the vibrations signal does not;
2) the vibration signal reflects raceway activity but the tuyere pressure does not
3) the standard signal analysis procedure is unable to detect true coherence due to contamination by higher frequency 'fine structure' effects.

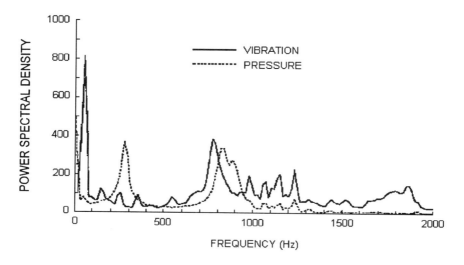

**Figure E.6** Comparison of tuyere 13 pressure and vibration power spectral densities.

Since the raceway turbulent energy has a short mixing length (or mean free path) this result is not surprising. The tuyere pressure in pneumatically coupled to the turbulent zone at the tuyere nozzle and it would have a low (attenuated) 'transmissibility', particularly at the higher frequencies where turbulent activity is more pronounced. The vibration signal, however, is structurally coupled to the turbulent zone in the tuyere nozzle, and it would have a high 'transmissibility'. This is confirmed by the tuyere punching results, which produced markedly increased vibration activity in the tuyere raceway.

In order to investigate the merits of the last conclusion, two further signal analysis methods were attempted, namely,

*a)* computation of the PSDs at 4 second time intervals and cross-correlating between pairs of tuyeres the movements of the individual peak amplitudes after digital filtering;

*b)* digitisation of the signals, narrow bandpass digital filtering (750-859 Hz), envelope detection and crosscorrelation between the conditioned time series data.

Neither method produced any significant cross-correlations. This indicates that there is no transfer function relationship between the two signals in a short time frame, and is a further indication that the vibration signals reflect raceway activity to a larger extent than the pressure signals.

### E.6.4    Tuyere-Tuyere Vibration Analysis

Azimuthal variations around the ISF were investigated by a multivariate model using the concepts in Chapter 7, in which vibration measurements at tuyeres 2 and 9 are the *inputs*, and the vibration measurement at tuyere 13 is the *output*. Ordinary coherence analysis between pairs of tuyeres showed there was little cross-talk between the tuyere raceways, at least for the data analysed in this study. Operating experience and the relatively short turbulent mixing length tend to confirm there is no interaction across the furnace, but it is possible that raceway interaction may occur between adjacent tuyeres. However, this was not part of the present investigation.

### E.6.5    Condenser Vibration Analysis

The accelerometers on the condenser off-take roof were unable to detect any change in turbulent noise activity caused by the build-up of a zinc curtain, primarily because

1) the signals were too highly contaminated by noise from the condenser rotors nearby, and

2) there was little zinc curtain build-up during the test period.

## E.7 CONCLUSIONS

The major conclusions that can be drawn from this investigation are as follows.

1) The vibration signals from the accelerometers mounted on the tuyere flanges, gave accurate measures of the turbulent/raceway activity in tuyere nozzle region. This is due to the efficient structural transmissibility of the vibrations along the tuyere piping for a broad frequency range.
2) The vibration spectral energy was most pronounced in the 6 to 9 kHz frequency bandwidth, and movements in the PSD peak amplitudes in this frequency range were highly correlated with gross changes in bustle main pressure and/or blast volume flow.
3) The tuyere pressure and vibration signals were uncorrelated for the broadband frequency range considered, thus indicating that tuyere pressure noise is not a true measure of raceway activity.
4) The vibration signals from tuyeres 2, 9 and 13 were uncorrelated when analysed using standard spectral analysis techniques. This indicates that there is no raceway interaction either across or along the ISF over extended distances in a short time scale.
5) The monitoring of raceway activity using vibration signals may enable a variety of metallurgical phenomena in the ISF to be investigated, including the buildup of speiss on the slag/lead interface, interaction between adjacent tuyeres, changes in slag level, height, etc.

## E.8. ACKNOWLEDGMENTS

The authors gratefully acknowledge the assistance of the management, metallurgists and operating staff at Sulphide Corporation Limited.

## REFERENCES

Romberg, T. M., & Harris, R. W., 1979, Application of Vibrations Signals to the Identification of Hydrodynamic Instabilities in a Heated Channel, *Journal of Sound & Vibration*, 65(3), 329-338.

Cybula , C, J., Harris, R. W., & Ledwidge, T. J., 1974, Location of Boiling Noise Source by Noise Analysis Techniques, *Proc. of the Institution of Radio & Electronic Engineers, Australia*, 35' 310-316.

Mucciardi, F., Stirring of Melts: A New Technique to Quantify Mixing, McGill University, Montreal (unpublished paper).

Harris, R. W., & Ledwidge, T. J., 1974, *Introduction to Noise Analysis*. Pion, London,.

Bendat, J. S., & Piersol, A. G., 1980, *Engineering Applications of Correlation and Spectral Analysis*. Wiley-Interscience, New York.

# Case Study F

## EXTRUSION DIE DYNAMICS

This case study is an edited version of a paper by T. M. Romberg, A. G. Cassar
and R. W. Harris

### F.1    CONTEXT

The structural vibrations monitored on a process plant often contain burst
phenomena where several discrete frequencies are present. Spectral analysis of
these vibration signals by Fourier transform methods is generally considered
incapable of resolving discrete frequencies close together in a short time slice.
The power spectral density estimates computed by traditional Fourier transform
methods are computed by more recent maximum entropy spectral analysis
(MESA) methods for both data from a test system where the outputs of two
tuned filters excited by pseudo-random binary noise are summed, and the
vibration signals from an accelerometer mounted on a die in an extrusion
process. The frequency resolution characteristics of the two methods are
compared, and the problems associated with the choice of an optimum order
autoregressive prediction error filter for the MESA method are discussed.

### F.2    INTRODUCTION

The power spectral density is one of the basic descriptors for 'signature
analysis' in the study of signals obtained from vibration transducers, and has
been very successful in determining the mechanical integrity of moving
machinery. The power spectral density is usually computed from digitised data
using a discrete Fourier transform (usually a fast Fourier transform) algorithm.
This method for calculating the power spectral density suffers from 'leakage'
caused by truncation of the time series record, and window functions have to be
used which are independent of the statistical properties of the signals being
analysed. The leakage problem becomes particularly acute for short data
records where the uncertainty principle based on the product of resolution
bandwidth and signal duration does not allow good frequency resolution.

Maximum entropy spectral analysis (MESA) techniques differ from the
linear discrete Fourier transform (DFT) methods in that they are data-dependent
and nonlinear, and hence do not assume periodic extension of the data as is

required in the DFT, or that the data outside the available record are zero to prevent 'leakage'.

The DFT and MESA methods for computing the power spectral densities are compared for both a simulated system of two resonant filters with marginally different natural frequencies, and data obtained from the vibration of a die assembly in a wire extrusion process.

The discrete Fourier transform (DFT) of a signal $x(t)$ sampled at $N$ equally spaced points at a time interval $\Delta t$ is defined by

$$X(f) = \Delta t \sum_{n=0}^{N-1} x_n e^{-2j\pi fn\Delta t}, \qquad \dots \text{(F.1)}$$

where $0 \le t \le T$ and $T = N\Delta t$. The unsmoothed power spectral density estimates are then calculated from the equation

$$S_{xx}(f) = \frac{1}{N\Delta t}\left[X^*(f)X(f)\right], \qquad \dots \text{(F.2)}$$

where $X^*(f)$ denotes the complex conjugate of $X(f)$. In order to minimise the leakage in the spectral estimates due to truncation in the time series, the estimates are smoothed, for example, by a Hanning window.

The MESA method computes the spectral estimates from the Burg algorithm after constructing an autoregressive prediction error filter (PEF).

$$S_{xx}(f) = \frac{P_M}{2f_s\left|1+\sum_{m=1}^{M} a_m \exp(-j2\pi mf\Delta t)\right|^2} \qquad \dots \text{(F.3)}$$

Equation (F.3) is derived by maximising the entropy (a measure of the information content) of the signal or, more precisely, the entropy rate with respect to the elements of the extended autocorrelation function

$$R_{xx}(m), \quad |m| \ge M + 1. \qquad \dots \text{(F.4)}$$

$P_M$ is the residual power of the $M$th order autoregressive PEF with coefficients $a_m$, and $f_s$ is one half the sampling frequency. The PEF coefficients may be evaluated by recursive, least squares and maximum likelihood computational algorithms. A review of MESA techniques is given in reference.

A major difficulty with the MESA method is the selection of the 'optimum' order filter (i.e. $M$ in equation (F.3)). If the order of the autoregressive PEF is too small, resonances in the data cannot be resolved, and the estimates will be biased. If the order of the PEF is too high, spurious peaks (instabilities) occur in the spectral estimates which result in a large variance. Many MESA algorithms determine the optimal order PEF by seeking a minimisation of Akaike's final prediction error criterion

$$FPE = \frac{N + M + 1}{N - M - 1} P_M ,$$                                    ... (F.5)

where $N$ is the number of data points and $M$ the filter order. (See Case Study A for the definitions of some other criteria.) The criteria attempt to give the best mean square compromise between bias and variance in the spectral estimates, and are good approximations only asymptotically (*i.e.* as the number of samples tends to infinity). Window resolution/window instability problems occur in calculating the spectral estimates by the DFT method. In this case, window closing/window carpentry procedures (such as weighting the data with a taper function at the ends of the sample) are used to determine the window with the 'optimum' resolution bandwidth.

## F.3        TEST SYSTEM RESULTS

Before analysing the vibration signal monitored on an extrusion die, the DFT and MESA methods were applied to simulated time series data in order to highlight frequency resolution, frequency shifting and spectral instability problems. The data $x_k$ were generated by passing pseudo-random binary noise $w_k$ through two resonant filters with general equations of the form

$$x_k = 2\alpha \cos(2\pi f_0) x_{k-1} - \alpha^2 x_{k-1}$$
$$+ (1-\alpha)(1+\alpha) w_k - \alpha(1-\alpha)(1+\alpha)\cos(2\pi_0) w_{k-1}$$      ... (F.6)

where $\alpha$ was chosen to be 0.99 and the natural frequencies ($f_0$) were chosen to be 0.25 and 0.27 Hz respectively. The pseudo random binary noise inputs were uncorrelated, and the outputs were summed to give the required time series. The data record consisted of 2500 points, of which 180 data points providing a signal with pronounced modulated characteristics were chosen, as shown in Figure F.1. Each 'packet' is about 60 seconds duration and Figure F.2 shows how the log normalised FPE varies with the number of filter coefficients for 60, 120, and 180 samples with Haykin's algorithm (recursive technique). The results obtained with the various MESA techniques are similar, so the Haykin algorithm was chosen as representative of this approach. A number of local minima are evident, and this highlights the problem of indiscriminate application of Akaike's (or any other) criterion for the reasons discussed above. For 60 and 120 samples the global minimum is four coefficients, whereas the corresponding normalised power spectral density estimates for the MESA and

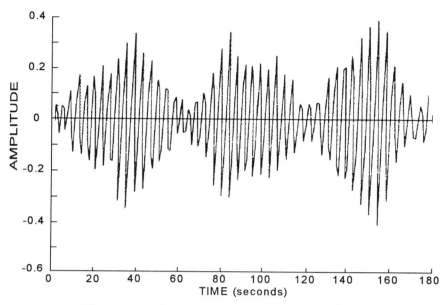

**Figure F.1** Simulated time series data (180 data points)

case with 180 samples has a global minimum at 21 coefficients. The DFT methods are compared in Figures F.3 and F.4. In Figure F.3 it can be seen that, with only 60 samples, neither the DFT nor the MESA method with four PEF coefficients was able to resolve the resonant frequencies (0.25 and 0.27 Hz),

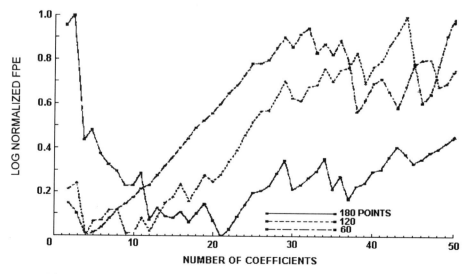

**Figure F.2** FPE for various sample lengths of simulated data (Haykin's algorithm)

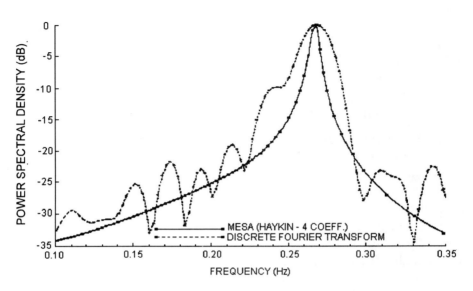

**Figure F.3** Power spectral density computed by DFT and MESA 60 points of simulated data

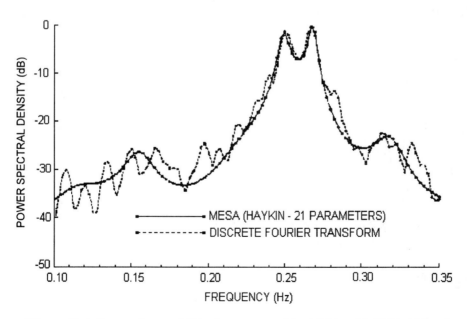

**Figure F.4** Power Spectral Density computed by DFT and MESA 180 points of simulated data.

but yielded a single resonance at 0.265 Hz. Figure F.4 shows that, with 180 samples, the resonant frequencies are resolved satisfactorily by both methods, with the MESA method with 21 PEF coefficients giving smoother spectral estimates. Note, however, that frequency 'shifting' has occurred with the MESA spectral estimates at the upper resonant frequency (*i.e.* 0.268 Hz as compared to 0.27 Hz).

## F.4 EXTRUSION VIBRATION RESULTS

The analysis procedures discussed in the previous section, were applied in a similar manner to the vibration signals monitored by an accelerometer mounted on a die assembly in an extrusion process for the manufacture of steel wire. The signals were recorded on magnetic tape and were digitised at a 25 microsecond sample rate to give a data record of 20,000 points. A 'slice' of 210 points (Figure F.5) which exhibited pronounced 'burst' phenomena was selected for detailed analysis. An FPE coefficient map for one of the leading 'bursts (54 samples) for all MESA algorithms is shown in Figure F.6. The Haykin and Andersen recursive algorithms yield coincident results with a global minimum at ten coefficients, whereas the least squares algorithms of Barrodale and Erickson and Marple have global minima at 36 and 35 coefficients respectively. The MESA power spectral density estimates with ten coefficients again gave comparable (albeit smoother) estimates compared to the DFT method, as shown in Figure F.7. When the complete slice of 210 samples was used all methods gave a global minimum of seven coefficients, as shown in Figure F.8. The corresponding estimates for this case are compared in Figure F.9. Both spectral results (Figures F.7 and F.9) show a dominant resonance at 4800 Hz.

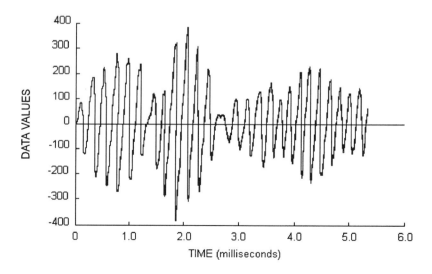

**Figure F.5** Vibration time series data

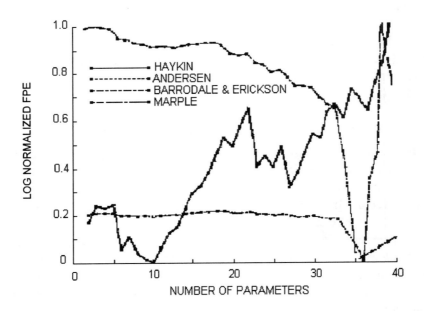

**Figure 6**  FPE for 54 points of vibration data using four separate MESA techniques

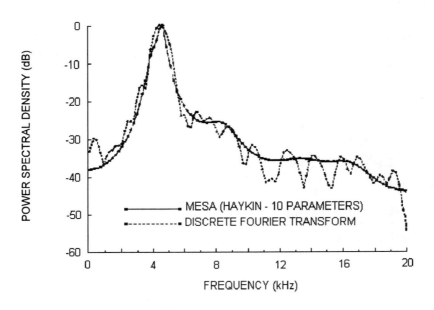

**Figure F7**  Power Spectral Density computed by DFT and MESA 54 points of vibration data.

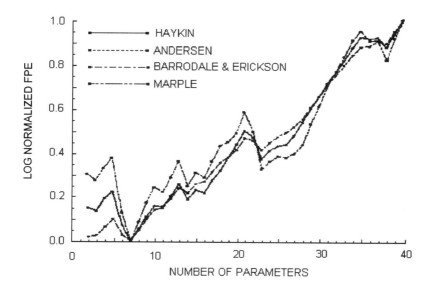

**Figure F.8** FPE for 210 points of vibration data using four separate MESA techniques

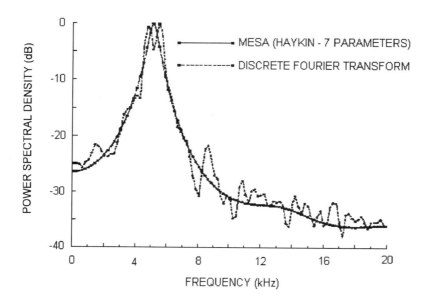

**Figure F9** Power Spectral Density computed by DFT and MESA 210 points of vibration data.

### F.5      CONCLUSIONS

Simulated data and actual vibration data were analysed to yield estimates of the power spectral density by both the MESA and DFT techniques for various sample lengths. In both cases, the MESA technique did not demonstrate superior frequency resolution over the DFT technique; however, the MESA estimates for the power spectral density gave smoother curves without the 'leakage' problems associated with the DFT, and also a narrower bandwidth for the resonance peak(s). These advantages are inherent in the parametric estimation properties of the MESA technique. The MESA technique requires an estimate of the number of coefficients required in a prediction error filter and this was shown to be a potential problem, particularly when using a least squares algorithm.

### F.6      REFERENCE

Romberg, T. M., Cassar, A. G., & Harris, R. W., 1984, A Comparison of Traditional Fourier and Maximum Entropy Spectral Methods for Vibration Analysis, *J. of Vibration, Acoustics, Stress and Reliability in Design*, Trans. of the ASME, **106**, January, 36–39.

# Case Study G

## MINERAL PROCESS DYNAMICS

This case study is an edited version of a paper by T. M. Romberg
and W. S. V. Jacobs

### G.1    CONTEXT

Australian mining companies have reported increased mineral recovery due to
better control of their concentrators based on information from their
radioisotope on-stream analysis (ROSA) probes. The ROSA probes and other
process instrumentation are usually interfaced with a mini-computer for data
acquisition and control of the plant. Various computer control schemes have
been developed to improve some aspects of plant operation, and these have
resulted in a decrease in mill overloads and a more consistent particle size in
the feed to the flotation circuits. Most flotation circuits, however, are controlled
manually from information obtained by regular laboratory assays or the on-
stream analysis of sample streams.

The New Broken Hill Consolidated Ltd (NBHC) collaborated with the
Australian Atomic Energy Commission (AAEC), now the Australian Nuclear
Science and Technology Organisation (ANSTO), in a joint study which was
aimed at improving the on-line control of their lead and zinc flotation circuits.
Correlation and spectral analysis of normal operating data provided by NBHC
enabled process relationships to be quantified, and demonstrated the feasibility
of using random fluctuations in the plant variables for process identification.
However, later analysis of some plant data indicated that it was influenced by
process and/or control feedback effects which are described in this case study.

### G.2    CLOSED LOOP PROCESS IDENTIFICATION

Most industrial processes have some form of closed loop control to ensure
stability of operation. The process input variations are consequently modified
by past output variations, and this complicates the identification of the process
dynamics when the input measurements are made within the closed loop. The
methods for identifying closed loop processes have been an active area of
research over the years, and readers are referred to Chapter 6 and the control
literature for further information.

In the work discussed in this case study, the closed loop identification was performed by modelling the cross-correlation estimates between the input and output time series. This has the advantage that, for stationary random processes, the feedforward and feedback contributions are readily discernible from, respectively, the positive and negative lag regions of the cross-correlograms. In addition the transport delays associated with the transfer of information through a distributed process are readily determined from the maxima in the cross-correlation estimates. The identifiabilty of closed loop processes using cross-correlation estimates is discussed in Chapter 6.

## G.3        PLANT AND DATA ACQUISITION SYSTEM

A simplified schematic arrangement of the NBHC plant is shown in Figure G.1. It consists essentially of primary and secondary grinding circuits followed (in turn) by the lead and zinc flotation circuits. The primary grinding stage has two rod mill and ball/rake classifier circuits in parallel followed by a primary lead flotation stage (approximately 70% recovery). The secondary grinding stage has a ball mill with two rake classifiers in parallel. Particle size of the feed to the lead and zinc flotation circuits is controlled by recirculating the oversize

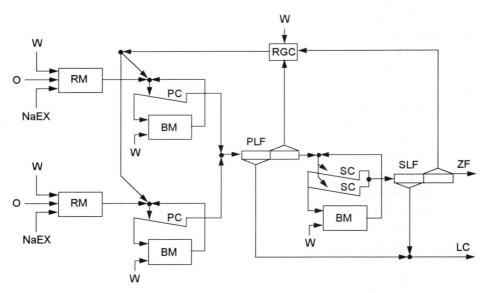

Legend: Ball Mill (BM); Lead concentrate (LC); Sodium ethyl xanthate addition (NaEX); Ore feed (O); Primary rake classifier (PC); Primary lead flotation (PLF); Rod mill (RM); Return to grinding circuit (RGC); Secondary rake classifier (SC); Secondary lead flotation (SLF); Water addition (W); Zinc flotation feed (ZF).

**Figure G.1** Simplified schematic diagram of the NBHC plant particles from the rake classifiers to the ball mill for regrinding. Details of the grinding circuit control are given in the reference.

Twenty-one streams of data were logged at half-minute intervals in the grinding and flotation circuits, which provided 40 hours of data in total. Only the process variables affecting the lead concentrations in the number one rougher tailings were investigated in detail. Correlation analysis of the data logged in the zinc flotation circuit indicated that it was contaminated and therefore unsuitable for modelling purposes.

## G.4     VARIABLES AFFECTING LEAD FLOTATION

Operational experience indicated that the lead concentration in the number one secondary rougher tailings (% lead Pb 1SRT), which is used to monitor the lead recovery in the flotation circuit, was mainly influenced by the sodium ethyl xanthate (NaEX) addition to the grinding circuit, the volume of slurry from the primary classifier overflows (PCOF), the lead concentration in the PCOF, and the particle size and slurry alkalinity of the secondary classifier overflow (SCOF). Their relative influence has been assessed in a previous analysis of the data, in which the feedback effects were assumed to be negligible. This feedback is principally associated with the return feeds from the flotation circuits to the primary grinding circuits.

In the investigation reported in this case study an assessment was made of the effects of the grinding circuit variables and the secondary circuit control action on the lead recovery in the flotation circuits, as monitored in the number one secondary rougher tailings (% Pb 1SRT). Bearing in mind the limitations associated with the direct application of spectral methods to closed loop processes discussed in Chapter 6, a preliminary indication of these effects was obtained from the partial coherence estimates of a multivariate model relating ore feed, PCOF lead concentration and secondary classifier motor current, of rake amp., as the inputs, and the lead concentration in the number one secondary rougher tailings as the output. The ore feed is outside the closed loop, whereas the other inputs are influenced by the return feeds to the grinding circuit, which marginally affect the partial cross-spectral density (hence coherence) estimates.

The input-output partial coherence estimates were computed by the recurrence algorithm from the augmented matrix of ordinary spectral density estimates (see Chapter 7), and are shown in Figure G.2. Note that of the grinding circuit variables, the PCOF lead concentration is most coherent with the output (% Pb 1SRT below 1.5 mHz, even though its 'energy' had to pass through the primary flotation and secondary grinding circuits as well as the first rougher bank of 12 cells in the secondary flotation circuit. The secondary classifier rake amp., a 'control' variable which is a measure of the recirculating

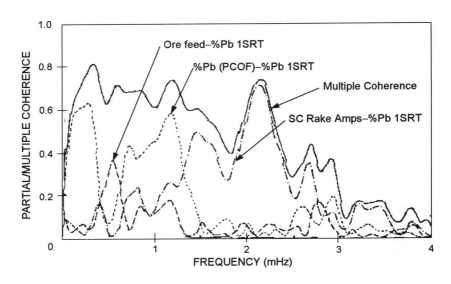

**Figure G.2** Partial and multiple coherence estimates

lead in the secondary circuit, has its coherent 'energy' transfer centred around a maximum (0.72) at a frequency around 2 mHz. Its relationship to the output is via the density of the SCOF, since a decrease in density is commensurate with a finer grind and better flotation characteristics of the lead particles. The relative insignificance effect of the ore feed on the output is indicative of the variations in lead content of the new ore. It should be noted that the time series data were passed through a recursive filter (bandpass 0.2 to 4.0 mHz) before computation of the partial spectral density estimates.

## G.5      PROCESS AND CONTROL MODELS

The partial coherence estimates suggest that the plant behaviour and control may be partially characterised by the following dynamic models.

- A 'process' model which relates the transfer of lead from the primary grinding to the secondary flotation circuits (denoted % Pb [PCOF]→% Pb 1SRT).
- A 'control' model which assesses the effect of the secondary grinding circuit recirculating loads (or rake amp.) on the SCOF density (denoted SCRA→SCOF density).
- A 'process' model which relates the SCOF density to the lead concentration in the number one rougher tailings stream (denoted SCOF density→% Pb 1SRT).

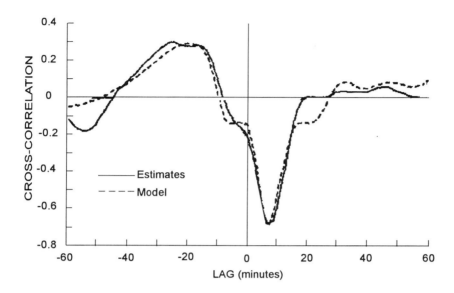

**Figure G.3** Cross-correlation between % lead (PCOF) and & Lead in No. 1
secondary rougher tailings

The cross-correlation estimates for each of these models are shown in Figures G.3 to G.5 respectively. The time series data for the process and control models were bandpass filtered in the frequency ranges 0.2 to 2.0 mHz and 0.2 to 4.0 mHz before computation of the cross-correlation estimates. These filter settings were chosen to pass the maximum pertinent information, and were determined from their ordinary cross-spectral density estimates.

The cross-correlation estimates for the 'process' model % Pb [PCOF]→% Pb 1SRT, Figure G.3, result in a minimum of −0.68 at 9 minutes lag. This response time was consistent with the operational experience but the negative minimum has proved difficult to interpret metallurgically. The estimates in the left half plane suggest a damped low frequency feedback which is probably associated with the return feeds to the primary grinding circuit.

The cross-correlation estimates for the 'control' model SCRA→SCOF density, Figure G.4, have a minimum of −0.55 at 1 minute lag, and indicate a highly oscillatory closed loop relationship between the variables owing to the secondary grinding circuits control algorithm (Whiten and Roberts(1974)). For a manually set ore feed rate, this algorithm controlled the water addition to the secondary classifiers to maintain a maximum circulating load concentration with the capacity of their mills. Thus an increase (decrease) in circulating load, which is measured qualitatively by the rake motor current, is accompanied by an increase (decrease) in classifier overflow density.

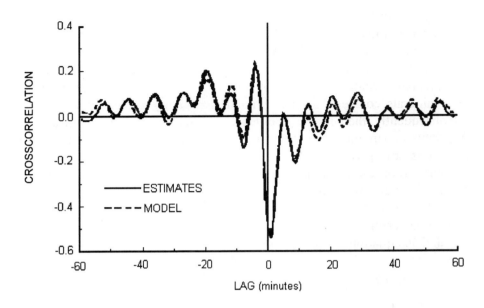

**Figure G.4** Cross-correlation between SC rake amps and SCOF density

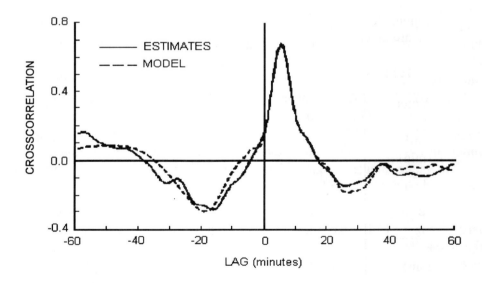

**Figure G.5** Cross-correlation between SCOF density and % lead in No. 1
secondary rougher tailings

The cross-correlation for the 'process' model SCOF density→% Pb 1SRT, Figure G.5 has a maximum at +0.69 at 5 minutes lag, which is also consistent with physical considerations. The increase in SCOF density was accompanied by an increase in the average particle size and, because the NBHC mill had barely adequate grinding capacity, this increase produced a greater proportion of lead minerals coarser than the optimum size range for maximum recovery rate. Consequently, the percentage of lead in the tailings stream increased.

The optimum (minimum least squares) models for these process and control relationships are given by the dotted lines in Figures G.3 to G.5. The parameters were estimated as in the test process, but with the estimation procedure included the transport delays and the parameters of second order models. The second order models gave some improvement to the minimum least squares errors; the improvement with higher order models was only marginal. The equations for a second order model are as follows:

$$R_{xy}(k) = \sum_{j=1}^{2} \left[ a_j \hat{R}_{xx}(k - d - j) + b_j \hat{R}_{xy}(k - j) \right] \quad k > k_0$$

$$R_{yx}(k) = R_{xy}(-k) \qquad\qquad k > k_0 \qquad \dots \text{(G.1)}$$

$$= \sum_{j=1}^{2} \left[ c_j \hat{R}_{yy}(k - f - j) + d_j \hat{R}_{yx}(k - j) \right]$$

The minimum least squares estimates of the feedforward and feedback parameters are given in Table G.1. The following points could be noted:

**Table G.1** Parameter Estimates of Process and Control Models

| Parameter | % Pb (PCOF)→ % Pb 1SRT | SCRA→ SCOF Density | SCOF Density→ % Pb 1SRT |
|---|---|---|---|
| Feedforward: | | | |
| Delay $d$ (min) | 9 | 0 | 2 |
| $b_1$ | −0.1199 | 0.5780 | 0.1810 |
| $b_2$ | 0.0100 | −0.7290 | 0.3490 |
| $-a_1$ | −0.9930 | −1.1610 | 1.2589 |
| $-a_2$ | −0.0750 | −0.9330 | −0.4120 |
| Feedback: | | | |
| Delay $f$ (min) | 12 | 0 | 15 |
| $d_1$ | 3.914 | −6.1418 | −0.0160 |
| $d_2$ | −2.5825 | 6.8238 | −0.0170 |
| $-c_1$ | 0.0520 | 0.0200 | −0.1670 |
| $-c_2$ | 0.9420 | 1.0230 | 0.9260 |

1) The delays were estimated first to ensure that the parameter estimates were minimum phase, and the starting values had to be in the right region for convergence of the estimate procedure; and
2) the feedforward and feedback cross-correlation estimates were modelled from the first zero crossing (lag $k_0$) to the left of the zero lag axis ($k=0$) in order to minimise discontinuities in the model starting values.

The overall agreement between the models is generally good. The errors in the feedback models for the % Pb (PCOF)→% Pb 1SRT and SCOF density→% Pb 1SRT relationships were slightly higher than their feedforward models because of the fundamental and higher mode characteristics of the process.

Stability analysis of these closed loop models requires that their characteristic equation

$$F(z) = \sum_{k=0}^{n} a_k z^k \qquad \qquad \text{... (G.2)}$$

satisfy the following constraints:

$$F(1) > 0$$
$$(-1)^n F(-1) > 0 \qquad \qquad \text{... (G.3)}$$
$$|\beta_k| < 1, \quad k = 1,2,3,\ldots,n-2$$

where, for the present models,

$$\alpha_0 = -a_2 c_2$$
$$\alpha_1 = -(a_1 c_2 + a_2 c_1)$$
$$\alpha_2 = -a_1 c_1$$
$$\alpha_3 \ldots \alpha_{n-5} = 0$$
$$\alpha_{n-4} = b_2 d_2 \qquad \qquad (n = d + f + 4) \qquad \qquad \text{... (G.4)}$$
$$\alpha_{n-3} = b_1 d_2 + b_2 d_1$$
$$\alpha_{n-2} = b_1 d_1 - b_2 - d_2$$
$$\alpha_{n-1} = -(b_1 + d_1)$$
$$\alpha_n = 1.$$

The coefficients $(\beta_k)$ are derived by the division method of the modified Schur-Cohn test from

$$\frac{F_k^*(z)}{F_k(z)} = \beta_k + \frac{F_{k+1}^*(z)}{F_{k+1}(z)}, \quad k = 0,1,\ldots,n-2 \qquad \ldots \text{(G.5)}$$

in which $F^*(z)$ is the reciprocal polynomial of $F(z)$, and is given by

$$F^*(z) = z^n F(z^{-1}) = \sum_{k=0}^{n} \alpha_{n-k} z^k \qquad \ldots \text{(G.6)}$$

Substitution of the parameter estimates from Table G.1 in these equations yields the following results:

1) % Pb (PCOF)→% Pb 1SRT:
   $F(1) = 0.1483 > 0$,
   $F(-1) = 0.7739$ not less than zero
2) SCRA→SCOF density:
   $F(1) = -0.0301$ not greater than zero,
   $F(-1) = 16.9437 > 0$;
3) SCOF density→% Pb 1SRT:
   $F(1) = 0.0544 > 0$,
   $F(-1) = 0.2484$ not less than zero.

These results show that all models do not satisfy the first two stability constraints, and are marginally unstable. This suggests the presence of inherent instabilities in the grinding/flotation processes, but these have not been noticed in the operation of the plant. It should also be noted that these results are affected by tolerances on the transport delays and parameter estimates, and should be confirmed by further studies using new data and more comprehensive multivariate models.

## G.6    CONCLUSION AND LIMITATIONS

This investigation has shown how the random fluctuations in the process variables of a commercial mineral concentrator, when monitored during normal production, may be used to identify some of its significant closed loop relationships. Optimum closed loop models were identified from cross-correlation estimates relating

1) the transfer of the fluctuations in lead concentrate from the primary grinding to secondary flotation circuits,
2) the motor current of the secondary grinding circuit rake classifiers to the secondary classifier overflow density, and
3) the secondary classifier overflow density to the lead concentration in the secondary rougher tailings stream.

Stability analysis of these discrete models showed that they were all marginally unstable, and indicated the presence of inherent instabilities in the grinding flotation processes of the plant. However, process instabilities have not been detected from operational experience on the plant, and since the stability results are affected by tolerances on the transport delays and parameter estimates, they need confirmation by new data and more comprehensive multivariate models.

It was hoped that these further interesting results could be supported by further tests on another lead/zinc concentrator owned by The Zinc Corporation, Limited (ZC). This plant has fewer return feeds from the flotation to the grinding circuit, and is generally considered to be a passive, well-controlled plant, principally because of the finer grind of the product in the feed to the flotation circuits. In this more recent ZC/AAEC study, 25 streams were monitored by the company in their lead and zinc flotation circuits, and included pseudo-random perturbation of the chemical additive rates (approximately ± 12% of mean rate), namely, the sodium ethyl xanthate addition to the lead circuit, and the copper sulphate and Z200 additions to the zinc circuit. No significant correlations could be found between the variables of the feed and tailing streams in the lead and zinc flotation circuits, nor between the perturbations in the chemical additive rates and the lead and zinc concentrations in their respective tailings streams. The latter results suggest chemical saturation of the pulp. It could be noted that, in contrast to Canadian experience in flotation control, no instabilities were detected in the zinc circuit as a consequence of variations in the copper sulphate addition rates.

This description of recent results has been included to emphasis the point that the identification methods outlined in this paper are strictly applicable to 'dynamically active' plant; that is, those with significant variations in the mineral concentrations in the ore feed, coupled grinding and flotation circuits, variable grind, etc. Although Canadian experience indicated that 'at present dynamic models can contribute very little to the practical design of flotation controls', the results presented in this paper indicate that the identification of dynamic models warrants wider application and evaluation in suitably equipped commercial mineral concentrators.

## REFERENCE

Romberg, T. M., & Jacobs, W. S. V., 1981, Identification of a Mineral Processing Plant from Normal Operating Data, *Automation in Mining, Mineral and Metal Processing*, Proceedings of the 1980 IFAC Symposium, Montreal, O'Shea, J., and Polis, M., (Eds.) Pergamon Press, Oxford, 275–282.

# Index